Construction Practices for Land Development

About the Author

Dewberry, headquartered in Fairfax, Virginia, is a fully integrated engineering and architecture firm operating in more than 50 locations throughout the United States. It consistently ranks among the top 45 design firms by *Engineering News-Record*, top 20 engineering/architecture firms by *Building Design + Construction*, and the top 5 engineering firms by the *Washington Business Journal*. Working in multiple federal, state and local, and commercial markets, Dewberry's services include site/civil engineering and surveying; transportation, transit, and ports and intermodal design; water, wastewater, and water resources engineering; architectural and interior design; environmental, coastal engineering and resilience services; emergency management and mitigation; full-service geospatial mapping and analysis; and alternative project delivery inclusive of design-build, public-private partnerships, and turnkey construction. The firm enjoys a reputation for quality, deep subject-matter expertise, community engagement, and putting the client first.

Construction Practices for Land Development

A Field Guide for Civil Engineers

Land Development Handbook Series

Dewberry

Editor-in-Chief: Sidney O. Dewberry, PE, LS

Editor: Claire M. White, PE, ENV SP

Mc Graw Hill

New York Chicago San Francisco Athens London
Madrid Mexico City Milan New Delhi
Singapore Sydney Toronto

Library of Congress Control Number: 2018955235

McGraw-Hill Education books are available at special quantity discounts to use as premiums and sales promotions, or for use in corporate training programs. To contact a representative please visit the Contact Us page at www.mhprofessional.com.

Construction Practices for Land Development: A Field Guide for Civil Engineers

1 2 3 4 5 6 7 8 9 LWI 23 22 21 20 19 18

ISBN 978-1-260-44077-5
MHID 1-260-44077-X

Sponsoring Editor
Lauren Poplawski

Editorial Supervisor
Donna M. Martone

Acquisitions Coordinator
Elizabeth Houde

Project Manager
Aishwarya Gosain,
Cenveo® Publisher Services

Copy Editor
Girish Sharma

Proofreader
Savitha Vanathi

Indexer
Anand Shekhar

Production Supervisor
Pamela A. Pelton

Composition
Cenveo Publisher Services

Art Director, Cover
Jeff Weeks

To the hardworking Dewberry employees
who dedicate their talent, energy, and passion to building amazing places.

Contents

Foreword

I've been using the Dewberry *Land Development Handbook* in my classes at Virginia Tech for many years and have always appreciated the care and dedication that Mr. Sidney Dewberry and his talented staff took in the preparation of the series of editions of the Handbook. As a long-time educator with significant professional experience, I have always emphasized the applied and pragmatic knowledge base as solid preparation for students as they begin their engineering career. I know that we each share the mission of helping the next generation of engineers with our knowledge and experience, whether it's in the classroom or in gathering knowledge in these wonderful books.

This book, *Construction Practices for Land Development*, is a book that I wish I had when I was a student. It gives an excellent overview for young engineers on the various ways in which the land development process can work in both the private and public sectors. It describes the teams involved in the process and their roles during the different phases of the project which is critical for a civil engineer to understand as they are frequently the hinge pin to making sure everything and everyone works smoothly. It further describes the various project delivery methods in use today and clarifies the often confusing pre-construction administration process. There are clear illustrations and even case studies. Then the book tackles basic cost estimating before getting into the phase of full construction administration and the roles of the teams during construction. Finally, the book discusses various construction techniques and the always difficult construction closeout issues.

Construction Practices for Land Development is a comprehensive book that will truly benefit the next generation of engineers as well as the land development and construction industry. It should be an integral part of any civil engineering curriculum and it most certainly is one that we will use from here forward. A sincere thank you to Mr. Dewberry, Claire White, and the talented contributors for their time, energy, and dedication to this project!

Dr. Randy Dymond, PE, F.ASCE, D.WRE
Professor of Civil & Environmental Engineering
Coordinator, Land Development Design Initiative (LDDI)
Virginia Tech
Blacksburg, VA
October, 2018

Preface

When this business was launched in 1956 the Land Development Planning, Engineering, and Surveying professions were largely viewed as a backwater branch of Civil Engineering and not respected as a legitimate engineering field. Other consultants looked down their noses at anyone engaged in this practice and felt it was not "real engineering." Since land development consulting was how I made a living, I strongly resented the notion. I felt then, and still feel now, that this is a very noble profession. It requires expertise in all branches of civil engineering including surveying, roadway design, grading, drainage, water systems, wastewater systems, dry utilities, and environmental science; as well as knowledge of the related fields including urban planning, landscaping, archaeology, and architecture. We also have a responsibility to understand the economics, schedule, and vision of a project. But more important than the experience gained as a land development consultant, is the product of our diligent labor: we work to improve our communities.

For this reason, I have devoted my career to elevating this profession to the level it deserves.

In the early days, infrastructure was often considered an inconvenience that reduced the overall budget of the project. Few regulations required adequate drainage, utilities, and other infrastructure to provide good, reliable access to housing, employment, healthcare, public transportation, and retail developments. We often clashed with our clients over these issues as we advocated for sustainable and resilient infrastructure that exceeded the minimum standards. Gradually, the localities mandated better infrastructure and improved environmental performance through enhanced standards and regulations. These requirements are still progressing and evolving today as evidenced by the tremendous strides taken in the Green Building and environmental movements. Today, most developers and localities acknowledge the importance of infrastructure investments.

The profession of Land Development Consulting is now recognized and respected among the engineering disciplines. Every major A/E consulting firm has a land development practice. It is taught in many colleges and universities and in some cases, as its own specialty track within the civil engineering program. Young people are aware of and attracted to the profession. They enter this field inspired, bringing with them new ideas, the most recent technology, and a youthful perspective on the world that challenges us "old-timers" to keep pace with the speed of learning, rise above convention and truly innovate for the benefit of our clients and our communities. I feel that our firm, in its way, has greatly contributed and remains attuned to this dynamic industry with the *Land Development Handbook* series.

The first *Land Development Handbook* began as a dream of mine many years ago. In the mid-eighties I committed to developing a literary resource that could be shared with the

Pictured left to right: Sid Dewberry, Jim Nealon, Dick Davis when the firm was known as Dewberry, Nealon & Davis. Leaning on Nealon's era 1935 airplane.

civil engineering industry. Rather naively, I thought a book was something you sat down to do and finished within a few weeks' time. How surprised I was to learn that it would take years. The first edition of this text, in fact, took 7 years from start to finish. When looking for interest from publishers, I was pleased that many were anxious to publish and distribute the book. We then entered into what has become one of the most treasured and unique business relationships I have formed over the years with a premier technical publisher, McGraw-Hill Publishing. After the first edition, McGraw-Hill told us they would want us to update the handbook every few years (if the book was successful). By their measures this book is a best seller for the industry and continues to serve as a great resource. I'm proud to see the book in the offices of our clients, in the hands of students, and on the shelves of other design firms.

This business and our communities have changed dramatically since the first edition was published. To evolve with these changes, we regularly update the book to capture changes in policies and identify ever-changing design processes. With this fourth edition we have also expanded the scope of the book to better represent the expanded scope of land development. New technologies, tightened economics, and more complex projects require a broader range of knowledge. This fourth edition of the *Land Development Handbook* has evolved into three books to focus on (1) business, (2) design, and (3) construction. Together, these texts are the *Land Development Handbook* series:

1. *Development of the Built Environment* is all about business and economics of public and private projects. This book is meant to improve understanding and communication between consultants and developers/owners to ensure greater success of projects.
2. *Land Development Handbook* continues to focus on both the process and the technical design of civil engineering. We also emphasize the importance of public, private, and community relationships and involvement. We can't design in a silo—everything is connected to the community and the environment.
3. *Construction Practices for Land Development* describes the construction and operations of a project. Our industry is quickly trending toward design/build processes. Design is influenced by construction and operation considerations, and the design team should be actively involved in the construction.

I want to personally thank everyone who contributed to the fourth edition. Having been through this process several times before, I know the success of this exciting Dewberry endeavor is due to the dedication of each team member. A new edition of a book is no small task, and the development of two new books is a monumental effort. This latest edition truly represents a corporate-wide effort, as nearly all our 52 offices have contributed in both large and small ways. This diverse corporate presence has yielded valuable insight and fresh perspective from across the country.

Our lead contributors, Cody Pennetti, Chris Guyan, Kat Grimsley, and Claire White have demonstrated tremendous dedication and passion toward a shared vision of creating a great resource for the industry. This team is unique with backgrounds in both professional consulting and academia that adds depth to the content of the texts.

- Cody Pennetti served as an editor and contributor for this edition of the *Land Development Handbook,* and capably handled the complex task of managing the production of the three book *Land Development Handbook* series. Cody began his career with Dewberry and has always been passionate about the industry and continuous teaching and learning. Cody is now pursuing his PhD at the University of Virginia with a goal of serving in academia while staying involved in land development consulting. Cody lent a unique perspective and new ideas in this undertaking and helped produce a great resource for both the professional and academic industry.
- Christopher Guyan was instrumental in the development of this edition of the *Land Development Handbook* and operated as both an editor and contributor. Chris joined Dewberry after earning his undergraduate degree in civil engineering from Penn State. He is currently working on his MS in Urban and Regional Planning at Virginia Tech. He has shown tremendous promise with his design work and his efforts on the *Land Development Handbook*. Chris has a strong focus on the planning and design associated with land development. This focus has helped shape some of the underlying themes in this fourth edition to emphasize the importance of planned development and the community.
- Dr. C. Kat Grimsley was the primary contributor, writer, and editor for *Development of the Built Environment*. This new text benefits from Kat's extensive knowledge of commercial real estate. She is the director of the Masters of the Real Estate Development program at George Mason University, has served at the U.S. Department of State managing an international development portfolio, and has private sector experience working on financial modeling and commercial transactions. Kat completed her doctorate at the University of Cambridge in the United Kingdom with a focus on tenure security and international property rights and has since been appointed as a NAIOP Distinguished Fellow and member of the UN Economic Commission for Europe's Real Estate Market Advisory Group under the Committee for Housing and Land Management.

- Claire White was the primary contributor for *Construction Practices for Land Development*. Claire earned her bachelor's and master's degrees from Virginia Tech and has experience working in the consulting field, starting as an intern with Dewberry. She has recently transitioned into a teaching role at Virginia Tech where she focuses on land development and real estate courses. From her consulting and academia roles, Claire has witnessed the critical role of engineers and the development team during the construction phase of a project. She has structured the text to help engineers improve design and further contribute to project success by anticipating construction processes.

Thank you all for your commitment to this endeavor! I also recognize that the efforts of these individuals are supported by family and friends that work behind the scenes. I want to sincerely thank those that are always there to support us while we work on these demanding projects. I have continued to hold the role of editor-in-chief for the fourth edition of the *Land Development Handbook* and the two new books in the series and have been proud to work with this team.

I also want to extend a special thank you to Dottie Spindle, my administrative assistant, who took care of the little things, the big things, and everything in-between so that I could focus on the things that truly matter to me, like these books and this company. Keeping me on schedule and on task is a challenge, but it is one she embraced with a smile for 33 years up until her recent retirement. Her replacement, Janice Spillan, a career executive assistant, has picked up the pieces seamlessly and continues to move the ball forward. Many thanks to her as well.

Peer reviewers are a critical component of our text. Those who think writing is difficult should try peer reviewing (or editing); balancing criticism with encouragement is a tall order. Our peer reviewers rose to this task under tight timeframes and across great distances. Their expertise was invaluable and that they were willing to lend it to this endeavor speaks highly of their commitment to Dewberry, to their practice, and to mentoring others.

Engineers, construction professionals, and surveyors within Dewberry have provided input and reviewed the final book. Specifically, Kurt Thompson not only reviewed large components of the text but helped assemble professionals within and outside of Dewberry to contribute. Every region within the United States has certain procedures to go through to obtain permits to do land development construction. Developers and public- or institutional-sponsored projects generally retain the services of local professionals who are familiar with local regulatory requirements, but the basic principles of design are universal.

While many professionals made significant contributions to *Construction Practices for Land Development*, three individuals in particular were instrumental in creating this great resource. Dave Speed, retired Vice President with William A. Hazel, Inc., has four decades of construction experience, collaborating with Dewberry on a multitude of projects in various sectors. He provided key insight to standard construction processes and the connection between design and construction. Kyle White serves as a project manager with Virginia Tech Capital Construction and Renovations. He understands not only the design engineer's responsibilities but also the key issues facing owners. Melissa Emory with Fairfax County Park Authority has experience in construction and as an owner of public projects. She is well versed in bidding and construction administration from both the contractor's and owner's perspective.

I would like to extend my sincere gratitude to all our clients for your continued support. Many of you have willingly offered components of your projects for inclusion in the text and we are happy to have your cooperation in this unique project. These projects would, of course, not be possible without the shared vision of our clients. In *Construction Practices for Land Development,* we introduced case studies based those prepared by Engineers Joint Contract Documents Committee (EJCDC) to highlight the complexity of construction and some common disputes between the owner, engineer, and/or contractor.

Two other individuals that have contributed greatly to this edition are Matt Pennetti and Dave Huh. Matt is an artist who developed hundreds of new illustrations and had the complex task of creating graphics of technically complex topics. Dave is a talented photographer at Dewberry who has an amazing eye for showcasing our projects in the best way possible. Our profession relies heavily on communication through graphics (plans, drawings, details), and these two professionals have contributed amazing works of art that complement the technical content of the books.

Craig Thomas, Dewberry's General Counsel, helped us to initiate this project with McGraw-Hill, has overseen all the contractual arrangements since we first published the *Land Development Handbook* many years ago, and has been a valuable legal resource throughout. Thank you, once again, for your support in this endeavor.

Finally, I want to express my deep regard for our partner in this effort, our publisher Lauren Poplawski of McGraw-Hill. She took over a large project and has been instrumental in the development of the new edition of the *Land Development Handbook* and the expansion of the *Land Development Handbook* series. Thank you for believing in us, for helping us elevate land development consulting as a profession and making one of my dreams—this book—come true, again!

In 1956, if you had told me that our six-person land development consulting company would grow into becoming one of the top 50 A&E companies in the United States, I would have thought you were nuts! I learned the hard way that real estate development was subject to the many ups and downs of the economy. For that reason, we sought early on to diversify our company into other facets of the A&E business. This diversification effort has been hugely successful for us, but land development continues to be one of the primary underpinnings of our practice. We love it and every new project continues to get the enthusiasm and professional care that we gave when we were first trying to get established. I would urge the thousands of small land development consultants throughout the United States to always consider the connection between the projects across geographies, design firms, the environment, and communities. It's necessary to diversify our skillset to adapt to industry changes and economics while also considering the big picture of our responsibilities as design professionals. And ultimately, deep down, we get supreme joy out of helping plan and build safe, healthy, financially feasible, sustainable, and beautiful places for people to learn, work, worship, shop, play, and live.

When I was born, 91 years ago, the horse and buggy had almost been completely replaced by the horseless carriage. Telephone, radios, and electricity were getting to be the norm. Television was just coming on the scene. The information technology era was being developed. Population in the United States had grown from 120 to over 328 million today. Air travel was beginning to be a serious competitor of rail, road, and water. The United States had been completely rebuilt. It is now a different place—totally new in just one lifetime. My first great grandchild, who is 18 months old, will see even more rapid changes with driverless cars and intelligent machines taking over our day-to-day activities, along with changes in healthcare. All these things will surely be a challenge to the engineers and scientists to provide solutions. Just in the next few years, many of these changes will have a huge impact on society, and how the engineers and scientists provide solutions for them will be amazing. Oh how I wish I were a young engineer facing these opportunities to provide solutions to the many challenges!

Sidney O. Dewberry, PE, LS
Chairman Emeritus, Dewberry
Editor-in-chief

CONTRIBUTORS

Primary Editors and Contributors:

Sidney O. Dewberry, PE, LS
Editor-in-chief, contributor

Claire M. White, PE
Editor, primary contributor

Cody A. Pennetti, PE
Managing Editor, contributor

Additional Contributors:

Melissa Emory, PE

Kyle W. White, PE

Matt A. Pennetti
Illustrator

Dave Huh
Photographer

Craig N. Thomas, General Counsel Dewberry
Legal advice and guidance

Jonathan C. Shoemaker
Legal Counsel

Aileen Heberer, Molly Johnson, Kimberly McVicker
Dewberry corporate information, cover art content

C. Kat Grimsley, PhD and Chris J. Guyan
Land Development Handbook series coordination

Olugbemi Aroke
Advanced Construction Roles

Behzad Esmaeili, PhD
Advanced Construction Roles

Matthew J. Martin, CCM
Construction Administration, Construction Techniques

Lawrence A. McDermott
Certificates of Occupancy and Post Construction Services
(*Land Development Handbook,* 3rd ed.)

Robert "Skip" Notte, PE
Project Delivery Methods, Preconstruction Administration, Construction Administration, Construction Closeout, Specifications

Mary Munson Rouse Root, LS
Construction Stakeout Surveys
(*Land Development Handbook,* 3rd ed.)

Leo Segal, PE
Construction Cost Estimating
(*Land Development Handbook,* 3rd ed.)

Christopher A. Simon, PE, CCM
Advanced Construction Roles

David Speed
Construction Techniques, Construction Equipment

J. Thomas Tanner, PE
Construction Cost Estimating
(*Land Development Handbook,* 3rd ed.)

George A. Wigfield, PLS
Certificates of Occupancy and Post Construction Services
(*Land Development Handbook,* 3rd ed.)

Reviewers:

James Barnhart, PE, CCM
Construction Techniques, Construction Closeout

Jesus H. Echevarria, LS
Construction Stakeout Survey

Bill Fissel, PE
Complete Review

Devin M. Keeler, PE
Construction Stakeout Survey

Anthony S. Pecci, PE
Advanced Construction Roles

Kurt R. Thompson, PE
Project Delivery Methods, Preconstruction Administration, Construction Administration, Advanced Construction Roles

J.I. "Jack" Vega, PE
Overview, Construction Cost Estimates

Mandy J. Wright
Bibliography

Construction Practices for Land Development

CHAPTER 1

OVERVIEW

1.1. INTRODUCTION

A civil engineer's responsibilities in a land development project do not end with site plan approval. The complexity of modern land development requires the civil engineer to assume a larger role in the construction phase of a project and continue to represent the developer through project completion. Regardless of the contract structure, stringent regulations and complex systems necessitate the civil engineer's expertise to assist the contractor in providing a quality-constructed project within budget and on schedule.

1.2. OVERVIEW OF THE CONSTRUCTION PROCESS

The construction process varies for each project, depending on the project type, site conditions, delivery method, and public or private ownership. However, in general, the civil engineer will prepare site plans and the architect will prepare building plans, both of which are reviewed and approved by the local jurisdiction. Approved site plans are commonly referred to as permit drawings, which allow the contractor and/or developer to obtain the required permits for the project. The contractor will use the plans to construct the proposed improvements, coordinating with the design team and developer as needed to resolve construction issues such as unforeseen site conditions, errors or omissions in the plans, and/or additional work requested by the developer. The developer and the local jurisdiction will review construction progress at substantial completion, which is a mutually agreed-upon milestone when the site and/or building are sufficiently finished to allow for operations. If the site and/or building are accepted after this walk-through, the local jurisdiction will issue a certificate of occupancy, which allows for use of the facility. The design and construction team will both assist in the turnover of the project to the developer, including establishing any operations and maintenance procedures.

The delivery method the developer selects determines the contractual arrangement with both the design and construction teams for their respective services. With the traditional design-bid-build delivery method, the developer contracts initially with the design team to develop detailed plans, then procures a contractor typically by bidding the design plans. Ideally, the site and building plans are approved by the local jurisdiction prior to bidding the project, but often due to schedule constraints, the developer will use unapproved plans for bidding. In contrast, for a design-build project, the developer will issue a request for proposals (RFP) for design-build teams based on a prepared project specification, often accompanied by bridging documents outlining how the project requirements are expected to be fulfilled. In this case, the developer will issue only one contract for design and construction services. The design-build proposal submitted to the developer includes partial design drawings developed by the design team and a cost estimate prepared by the contractor. Once selected, the design team will finalize the drawings, and select construction activities may overlap the design phase.

The developer may also choose to use the construction manager at risk (CM at risk) delivery method, which maintains separate contracts for the design and construction teams but benefits the developer by promoting collaboration between the design team and contractor during the planning phases. The developer will have two separate contracts with the contractor—one for preconstruction services including estimates and constructability reviews and a second for the actual construction.

Both private and public developers manage land development projects; however, public projects have more regulated legal procedures as it is typically publicly funded. For public design-bid-build projects, the design team is selected through a qualifications-based selection (QBS) process, and the construction team is selected by low bid. Delivery methods used on other public projects will use a QBS process for both the design and construction teams. On the other hand, private developers have greater autonomy in assembling a development team for a project. They can choose any delivery method and any design or construction team, including sole sourcing the work to one firm. Often, private developers will solicit proposals or bids from multiple design and construction firms. They can acquire bids from specific firms or issue a public advertisement and are not bound to selecting the lowest bid.

Land development projects cover a variety of project types, including residential, commercial, industrial, and institutional. They can be a greenfield or redevelopment site, also referred to as a grayfield or brownfield. The project type will affect the sequencing of construction activities to address critical path milestones in order to deliver the project on schedule. Redevelopment of existing sites may present challenges through building demolition such as potentially hazardous materials, unmarked underground utilities, and maintaining access and utility service to adjacent developments. On the other hand, greenfield development may require extensive earthwork and infrastructure improvements to support the proposed development.

1.3. THE DEVELOPMENT TEAM

All projects begin with the developer, who establishes the need for the project and assembles a development team to achieve the project goals (See Fig. 1.1.). The development team, as defined in this text, includes all members involved in the project from pre-design efforts through construction. In general, the developer is the person or organization who contracts with the legal, business, design, and construction teams who work together to meet project requirements and develop a successful project. The developer may represent a private company or a government agency, and the developer may also be referred to as the owner or client. Developers associated with a government agency such as the school board or parks and recreation department are typically the owner of the property during design, construction, and through occupancy including operations and maintenance. A private developer, however, may or may not be the property owner at any point during the project and is purely responsible for bringing the project to market. Some private developers do retain ownership of the property following construction and operate the facility for an extended period.

The legal team includes land use attorneys who are primarily involved in the pre-design and design phases of the project to navigate land use and zoning laws. They will assist the team on legal aspects of development including deeds associated with easements, interpretations of local codes, document recordation, and authoring proffers or development conditions. Attorneys will have a major role at the beginning of projects that require rezoning, zoning variances, or special permits. They may serve as an advocate for the developer during negotiations with the public agencies and may lead presentations at public meetings when a project requires continuous public communication and updates from pre-design through construction. The involvement of the land use attorney will extend through the design process as they coordinate with the title company and into construction, working with the lender for construction draws to pay the contractor. The legal team is also involved with reviewing lease or sales contracts during and after construction. They may also assist with reviewing zoning and other legal

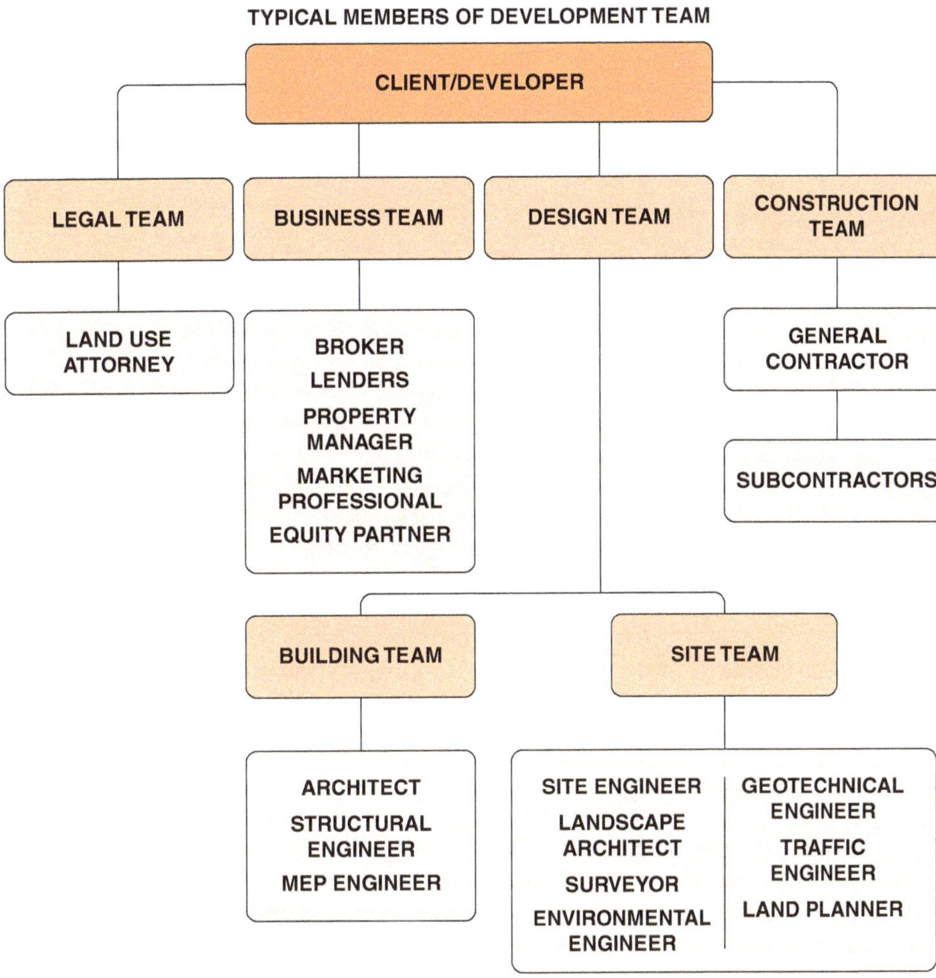

FIGURE 1.1 The development team.

components of the project for any design revisions during construction.

The developer will interact with the business team, starting in the due diligence phase and continuing through construction and opening of the project. A developer may work with an equity partner to share in the financial risk of the project. (See Fig. 1.2) After securing the financing from a lender in the due diligence and/or design phase, the developer will coordinate with the lender throughout construction to process draws on the construction loan to pay the contractor. Depending on the project type, the end user may be secured prior to construction commencement. However, in the case of speculative commercial properties, a developer will rely on a broker to secure leases for unoccupied space during construction and following building opening until all spaces are leased. A marketing professional will also advise the developer of potential tenants, appropriate sale or lease prices, demographics, and other geographical market conditions. Finally, a property manager will provide input on the operations and maintenance of the facility starting in the design phase and ultimately, serve as the developer's representative of the built facility.

The design team consists of both the building team, led by the architect, and the site team, led by the civil engineer. The design team members are required to be licensed professionals including architects, engineers, and surveyors to sign and seal plans that are submitted for permit. Their responsibilities include not only the development of construction drawings but also construction administration activities as further described in this book. The design team will contract directly

FIGURE 1.2 Construction activity for the Innovation Center, a mixed-use development adjacent to a planned Silver Line Metro station in Herndon, VA. This project is a joint venture between private developers and Fairfax County.

with the developer unless the design-build delivery method is used. With design-build, the design and construction team combined have one contract with the developer. Each project differs in its contract structure and in the quantity of contracts the developer has with the design team representatives. The developer may choose to have one contract with the design team, and typically, the architect administers the contract. In this scenario, the site team, including the civil engineer, is a subconsultant to the architect and plays a supporting role during construction administration. For other projects, a developer will have a separate contract with the building and site team, and each contract will often include both design and construction administration services. Regardless of the contract structure, each consultant on the design team will typically have a project manager responsible for coordinating with the developer, other design team members, and the construction team.

The general contractor typically leads the construction team and contracts directly with the developer. The general contractor provides oversight and management of the subcontractors such as the demolition, site work, and building trades. For single-family residential developments, the developer may contract directly with a site work contractor, not a general contractor, since the homebuilder will complete the building construction. The general contractor will have a superintendent overseeing the project, coordinating the trades, and representing the contractor daily onsite. The site work contractor will typically hire a surveyor to perform the construction stakeout survey stakeout and prepare the as-built survey if there is not a surveyor on the contractor's staff.

In addition to the design and construction team, the developer may choose to contract with other entities during construction. This may include a special inspector, who is a geotechnical engineer responsible for certifying structural elements, including subgrade, soils and foundations, and vertical elements. Special inspections are required for building pads, retaining walls, and parking or roadway areas as specified in local building and site codes. The special inspector may or may not be the same geotechnical engineer who performed the original subsurface investigation during the design phase. Material testing and quality control and assurance are required by the agency responsible for the infrastructure, and the developer must submit testing records at construction completion. In addition to special inspections, developers may also choose to hire a third-party inspector to evaluate private infrastructure separate from the public inspector assigned to the project. This inspector may serve as a commissioning agent to assist with testing all building systems prior to accepting the project from the contractor. Inspectors are involved throughout construction, regularly reporting to the developer and/or local jurisdiction on the status and quality of construction. Developers may also hire a construction manager (CM) as their third-party representative onsite. This may be particularly helpful for a developer who does not have in-house construction managers. The CM is typically onsite daily, provides reports to the developer, and aids in settling disputes. The CM is different from the general contractor's superintendent and the design team representatives, who may make regular site visits and observations.

The local jurisdiction that approved the permit drawings is also involved during construction, most notably with required site and building inspections. Typical site inspections include erosion and sediment control, stormwater, and utilities. These inspections are required to maintain construction permits, release bonds at the end of construction, and ultimately acquire the certificate of occupancy to open the project. The local jurisdiction's review staff will also be involved with preconstruction meetings and any required reviews of site plan revisions to address construction issues. In addition to the local jurisdiction's inspectors, the developer may contract with a third-party inspector to ensure the quality of the built project.

All members of the development team play a critical role in the success of the project. Constant coordination is required to quickly resolve any construction issues and help the contractor maintain the established schedule. Regularly scheduled progress meetings between the developer (or developer's representatives), construction team, and design team are a common way to provide updates on the construction status and upcoming action items.

1.4. THE CIVIL ENGINEER'S ROLE

The civil engineer is also referred to as the engineer of record of the approved site plans which detail erosion and sediment control, horizontal layout, water and sanitary sewer, grading, stormwater management, and landscaping. Civil engineers are involved in making decisions pertaining to the site and ensuring that the design is in adherence with the development constraints. Their knowledge of planning decisions during the design phase make their collaboration with the contractor during construction critical to the overall success of the project. The formal involvement of the civil engineer will vary based on the contracted scope for construction services with the developer. However, it is an asset to the developer for the civil engineer to be involved beyond merely implementing plan revisions to address construction issues.

The civil engineer may help with compiling documents to assist the developer with bidding the project. In addition to finalizing the site plans, the civil engineer's role in bidding includes preparing specifications, instructions to bidders, bid forms, and conditions. The civil engineer may also be asked to prepare a construction cost estimate based on the permit drawings for comparison with contractors' bids, as well as to be used for specific elements of site work that a jurisdiction can use to develop bond amounts. Once construction commences, if a developer does not hire a CM, the civil engineer may assume more construction administration responsibilities including coordinating with the contractor and observing construction progress. Even with a CM involved, the civil engineer will review shop drawings and requests for information (RFI) and may also be asked by the developer to participate in the final walk-through and punch-list development. Additionally, the local jurisdiction may require maintenance agreements for some site components and record drawings showing constructed conditions, both of which the civil engineer will assist in preparing.

A civil engineer may also be involved with construction as an inspector or CM hired by the developer. An inspector or CM should not be the same civil engineer who prepares the design documents and performs construction administration tasks. Typically a civil engineer serving as either the inspector or CM has training and specializes in these services. The inspector and/or CM may work for the same firm as the civil engineer or may work at a firm specializing in these services.

Regardless of the assigned responsibilities, it is important that the civil engineer develop a relationship with the contractor and that they work together to achieve solutions. The civil engineer must be willing to discuss the design and be receptive to revisions, particularly when the revisions save time and/or money while still meeting the required standards and specifications. Additionally, the civil engineer is responsible for knowing the construction documents, including the site plans, specifications, conditions, and contract. Within each of these documents are standard procedures that should be followed in an effort to remove emotions from contentious or debated decisions. Ultimately, the most successful projects for the developer occur when all development team members collaborate to achieve the project goals.

1.5. SAFETY ONSITE

As discussed in this book, a civil engineer will typically need to make a site visit to an active construction site. Though civil engineers do not visit the site as an Occupational Safety and Health Administration (OSHA) safety inspector, there are certain precautions and guidelines they should follow for their own personal safety. Prior to visiting a site, a civil engineer should notify the general contractor about their visit and confirm if any training is required to be onsite. For example, training may be required to access railroad property, and background checks and/or clearance approval may be required for government facilities or marine terminals.

Dress for site visits should include the following:

- Long pants
- Shirt with a minimum 4-inch sleeve
- Leather work boots (steel-toed boots or dielectric safety toed boots may be required)

If visiting a gas, oil, or electrical transmission or substation project, fire-resistant outer clothing with cotton inner clothing is required.

Personal protective equipment (PPE) is also required in addition to the dress outlined above. (See Fig. 1.3.) Each site condition may necessitate different PPE, but at a minimum, a hard hat, safety glasses, and safety vest are recommended on every site. The contractor may also require the civil engineer to wear cut-resistant gloves and hearing protection.

The civil engineer should also prepare for any outside hazards anticipated onsite. This includes applying sunscreen (SPF 50 or 70) to protect exposed skin and applying insect repellent if walking through tall grass, weeds, or wooded areas. It is also a good practice to check for ticks before leaving the site as well as when returning to the office or home.

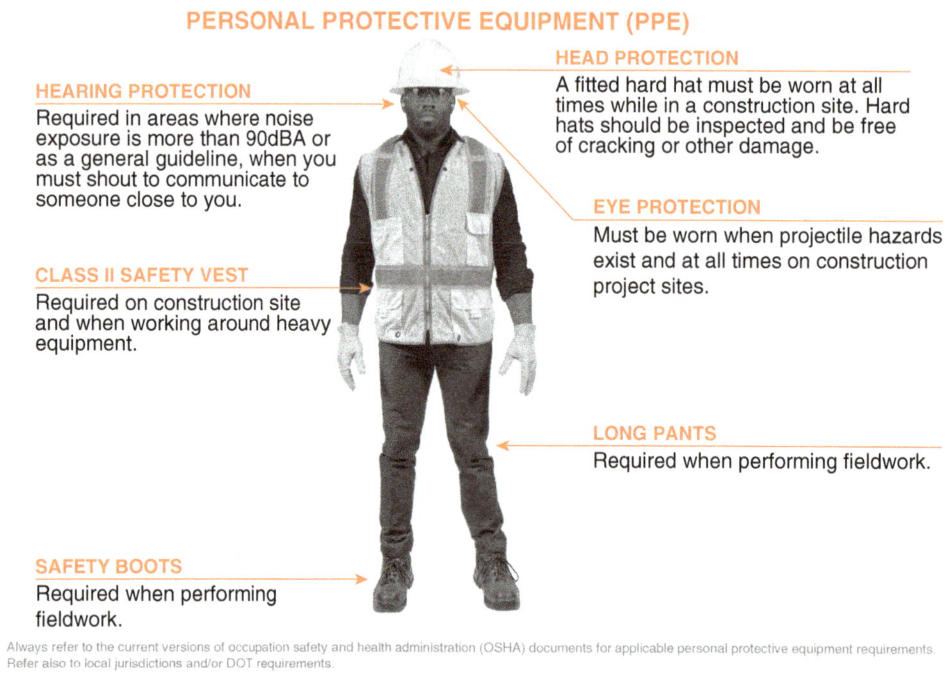

FIGURE 1.3 Personal protective equipment (PPE) required on a construction site.

When arriving onsite, the civil engineer should report to the general contractor's office to complete any required safety orientation and/or get an escort if required. At a minimum, the civil engineer should know the following:

- Whom to call in event of an emergency
- Where the rally point or meeting location is
- The location of the nearest first aid kit and the nearest medical facility

While walking around the construction site, the following safety guidelines should be observed by the civil engineer:

- General guidance
  - Look where you are walking and watch for uneven terrain, loose material, trash, slippery, or icy conditions.
  - While climbing ladders, maintain three points of contact at all times while climbing.
  - Do not ride in the back of a pickup truck unless it is designed with seatbelts.
  - Always wear a seatbelt while traveling in motorized equipment on the construction site.
  - Do not look at the welder's arc during welding operations.
  - Keep out of areas where chemicals are being used such as epoxy paints.
- Around the site
  - Stay alert and watch for equipment, pedestrian traffic, holes, surface water, etc.
  - Watch for wildlife and bugs including spider webs, yellow jacket nests, snakes, and alligators.
  - Never walk under overhead work or scaffolding.
  - Make eye contact prior to walking past heavy equipment.
  - Listen for backup alarms and avoid walking behind equipment in the operator's blind spot.
  - Walk around plywood flat on the ground covering holes.
  - Maintain 6 feet from the top of excavations or trenches deeper than 6 feet unless protected.
  - Access trenches deeper than 4 feet with a ladder or ramp.
  - Never enter a manhole or enclosed area without proper training and measures including a portable blower to ventilate the manhole, a 4-gas air monitor, appropriate harness equipment, and another person as a spotter depending on the type of risk.
  - Stay upwind of jack hammering and concrete sawing.
  - For any roadway work, stay behind jersey walls, cones, or barrels and remain alert of oncoming traffic.
  - Maintain as much distance as possible from high noise areas or use hearing protection such as ear plugs.
  - Never stand directly in front of a concrete pile being driven into the earth.
  - Never stand behind a tugger pulling electrical cable.
  - Never lean on a guardrail system (especially cables) or sit or lean on jersey walls.
  - Never walk under a load on a crane and do not touch the crane when working near power lines.
- In the building
  - Stay 15 feet away from the edge of the roof.
  - Stay back from open balconies or window frames where the window sill is less than 39 inches from the floor.
  - Do not access floors where concrete grinding or drywall sanding is occurring.

Lastly, prior to leaving the site, report any accidents or unsafe conditions to the general contractor.

CASE STUDY: STANDARD OF CARE

ISSUE:

The civil engineers' liability for failure of a rain tank that they designed, including issues of reliance on the manufacturer's information regarding the tank, standard of care, and contractor's responsibility.

SUMMARY:

First Baptist Church retained ABC Engineering to conduct various civil engineering tasks associated with the site of a new church. ABC Engineering designed a rain tank that would be located 10 feet under the new parking lot—the tank would be an initial collection point for storm water. As the project proceeded, the contractor raised concerns about the suitability of the rain tank, noting that the local water table was high, and posing questions about the details of installation. ABC Engineering did not reevaluate the rain tank, or provide any additional information about installation, but did provide assurances to the contractor that the ground water level would not affect the rain tank. Several months after installation, the tank and the parking lot above it collapsed, resulting in a substantial rework, and delays in occupancy of the church.

In subsequent litigation, the contractor and the owner offered expert testimony that ABC Engineering had breached the professional standard of care. In reply, ABC Engineering's experts contended that the failure of the rain tank was the result of construction errors by the contractor, and that ABC Engineering had met its standard of care by relying on the tank manufacturer's product information. The trial court concluded that the contractor had adhered to ABC Engineering's design, and that the sole cause of the collapse was ABC Engineering's failure to meet the professional standard of care. ABC Engineering appealed.

DECISION:

The Supreme Court of Virginia affirmed the lower court's ruling against ABC Engineering. (The court did decide in ABC Engineering's favor with respect to certain damages issues, and remanded the case for consideration of the impact of the owner's settlement of a claim against an inspection firm.) The court examined the record and found ample basis for affirming that ABC Engineering had not met the professional standard of care.

The court noted testimony that the specifications regarding the rain tank were prescriptive, and the contractor's duty was to follow them—refuting the notion that the construction contract had shifted design liability or duties to the contractor. Expert testimony in the trial court had also been critical of ABC Engineering's reliance on manufacturer's generic literature, and failure to conduct due diligence on the tank's location and the impact of the high water table. The design was labeled "not clear, constructible, or very likely to serve its purpose," apparently with respect to lack of direction regarding the need for a very level tank base and "nearly perfectly perpendicular" vertical panels. The engineer's failure to answer the contractor's question regarding installation was criticized.

The decision cites testimony from the contractor's and owner's experts favorably, but the point seems to be that there are factual grounds for affirming the lower court—the intent does not seem to be to establish broad precedential rules of law regarding what is and what is not within the standard of care. For example, in general support of the lower court's finding that the engineer did not meet the standard of care, the court cites testimony by an expert that the standard of care requires an engineer "to reexamine its original plan when the contractor submits an RFI about the suitability and performance of a structure." That may have been a valid point in the context of the church project, but perhaps was not intended by the Virginia high court as a new rule of law for design professionals. As another example, the court reported that an expert had opined that "an engineer that adopts the general plans and specifications prepared by the non-engineer manufacturer falls below the standard of care." It would seem reasonable to take this as a project-specific statement regarding ABC Engineering's liability, rather than a holding that has absolute application to all claims against design professionals; in some situations, a design professional's reliance on third-party information may be routine and entirely justified.

COMMENT:

The court here correctly employed a professional standard of care analysis in determining ABC Engineering's liability. That liability was strongly supported by expert testimony offered by the owner's and contractor's witnesses at trial.

Case study based on William H. Gordon Associates, Inc., v. Heritage Fellowship Church. *Supreme Court of Virginia (2016).*

The information provided does not constitute legal advice. Consultation with an attorney is strongly advised.

1.6. THIS HANDBOOK

This book approaches the construction stage of a project primarily from the vantage point of the civil engineer who prepared the design documents and explores the roles and responsibilities of the civil engineer during the bidding and construction processes. Construction terminology and processes commonly applied to land development projects are also introduced. However, many of the construction processes described are also applicable to civil engineers focusing on infrastructure design such as wastewater, water distribution, and transportation. While some construction processes such as bidding are more common with public projects, this book covers a variety of project types and highlights variability in the construction procedures based on the project delivery method.

This book is a reference for both students and professionals in the civil engineering, architecture, and land development industry, who possess a basic understanding of the development process, civil engineering terminology, design components, and construction methods. Readers should also have experience in reading civil engineering drawings, as discussed in the *Land Development Handbook*, either from coursework or prior work experience.

CHAPTER 2

PROJECT DELIVERY METHODS

2.1. INTRODUCTION

For every project, a developer will have contracts for design services and construction, and these contracts are part of the project delivery method. The American Institute of Architects (AIA) and Associated General Contractors (AGC) define a project delivery method as "…the method for assigning responsibility to an organization or an individual for providing design and construction services" (Cox et al. 2011, 2). This differs from project management which is "… the means for coordinating the process of design and construction" (Cox et al. 2011, 2). Project constraints such as schedule, size, complexity, and funding as well as a developer's preferences will dictate which project delivery method is used. A typical example is a developer preferring a single contract with the design team and separate contract with a general contractor (GC). In this case, the developer and the prime design consultant have a contract, and the prime design consultant has contracts with the other design consultants, also known as subconsultants. The developer's contract with the GC will ensure all construction activities, including the site work, fall under a single entity who coordinates all related construction activities. The GC will usually self-perform major portions of the construction while having many subcontractors and suppliers working under them to provide the full range of site and building construction.

When using a single contract for the design team, the developer typically contracts with the architect as the prime consultant, and the subconsultants may include the civil engineer, mechanical/electrical/plumbing (MEP) engineer, surveyor, and others. Alternatively, the developer may prefer to contract the land use and site-related design services under a civil engineer early on and contract the building design team separately under a selected architect. Each prime design consultant contracts with the necessary subconsultants and coordinates and manages their efforts effectively throughout all phases of the project. In this case, the civil engineer and the architect each have contracts with the developer and coordinate with each other in a cost-effective and timely manner that serves the interests of all parties.

The developer's legal counsel will prepare design and construction contracts or utilize standard contract documents from organizations such as the American Institute of Architects (AIA), the Engineers Joint Contract Documents Committee (EJCDC), or Design-Build Institute of America (DBIA). A developer can purchase these standard documents and edit them as needed to tailor them to a specific project. As an example, when developers use an AIA standard contract as the basis for a contract with a civil engineer, they can modify any references to the "architect" as the designer to "civil engineer," or it is assumed references to the "architect" refer to the "civil engineer." As professional organizations representing architects and engineers respectively draft these standard contracts, they inherently protect the architect's or engineer's interests. Therefore, a developer will negotiate the contract language to better balance the responsibility and risk shared among the contracting parties. The EJCDC, AIA, and DBIA contracts are available for purchase online.

Regardless of whether or not a developer uses a standard contract form, the contract defines responsibilities, payment, and schedule. As an example, contracts as well as drawings often designate the contractor is responsible for field verification of the survey and design and coordination of any discrepancies with the design team. A contract may also identify when payments are made to the contractor, any fees incurred for delays, and any incentives for an early project completion. Below is a list of items the civil engineer should consider when reviewing a contract for site design and construction administration services:

- Read the contract and understand its terms
- Prepare a defined scope of work, state exclusions, and provide appropriate design fees for scope
- Communicate expectations through the contract and adhere to these procedures during the design and construction phase
- Get input from within the civil engineering firm if asked to do something atypical rather than relying on a developer or contractor to indicate what is appropriate
- Carefully review modifications to standard contract language that shift more responsibility to the civil engineer
- Review insurance requirements and look for uninsurable terms such as indemnity provisions
- Review standard of care statements
- Understand obligation to continue work in absence of payment from the developer
- Review the ownership clause of intellectual property and include appropriate disclaimers if the developer will own design and therefore, may reuse it in the future
- Remove any statements where the civil engineer is responsible for safety onsite

As stated in the list above, the civil engineer should carefully review any standard of care statements in the contract. EJCDC includes the following definition of standard of care in their standard contract documents:

> The standard of care for all professional engineering and related services performed or furnished by Engineer under this Agreement will be the care and skill ordinarily used by members of the subject profession practicing under similar circumstances at the same time and in the same locality. Engineer makes no warranties, express or implied, under this Agreement or otherwise, in connection with any services performed or furnished by Engineer. (2014, 3–4)

When developers modify a standard contract or prepares their own contract, the civil engineer should carefully review any references to the standard of care equating to the best design in the industry, perfection, and warranties or guarantees of the work. It is common practice for civil engineers to strike such language and suggest alternative language that meets the standard of care that is customary and is covered under their professional liability insurance. Standard of care is used to determine negligence and therefore should be presented accurately in the contract.

2.2. PROJECT DELIVERY METHODS

A variety of project delivery methods are used for land development projects including:

- Design-Bid-Build
- Design-Build
- Construction Manager at Risk
- Public-Private Partnership

While developers may incorporate various adaptations to each delivery method, the basic framework of each delivery method is outlined in this chapter.

2.2.1. Design-Bid-Build

Design-bid-build is the traditional project delivery method where a developer has a contract for the project design and a second contract for the

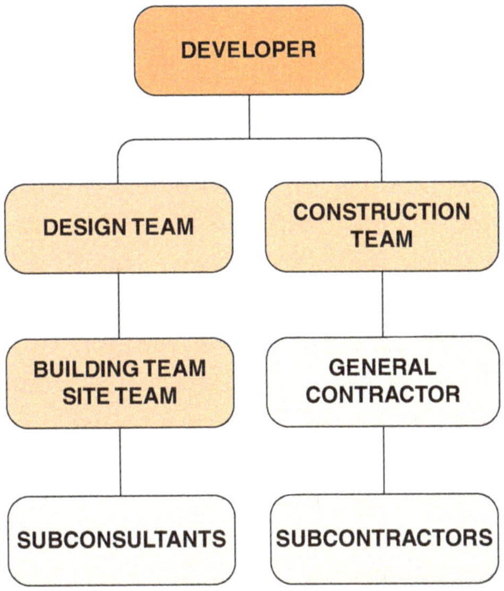

Figure 2.1 Design-bid-build project delivery method where the developer contracts with each design consultant individually and a general contractor who has separate contracts with subcontractors. An architect typically contracts with subconsultants such as structural, mechanical, and electrical engineers, and a civil engineer may contract with a surveyor.

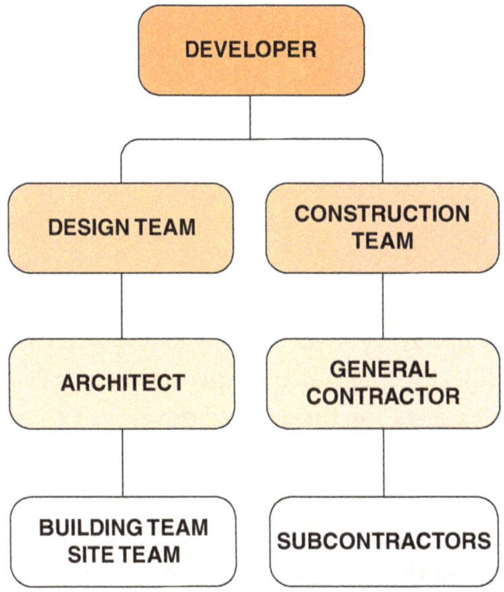

Figure 2.2 Design-bid-build project delivery method where the developer contracts with a prime consultant (architect shown above) and a general contractor who has separate contracts with subcontractors. The architect contracts with a civil engineer, who in turn typically contracts with a surveyor.

construction. For the design contract, the developer can contract with the building team (led by the architect) and the site team (led by the civil engineer) separately as shown in Fig. 2.1.

Alternatively, the developer can select one consultant (typically the architect) to be the prime consultant who has subconsultant agreements with other consultants on the design team as shown in Fig. 2.2.

In either scenario, the construction contract will be with a GC who will have subcontractor agreements with the different trades such as site work, building foundations, and masonry. Typically, the developer selects the design team based on a qualifications-based selection (QBS) process whereas the contractor is selected based on the lowest bid as shown in Fig. 2.3. As discussed further in Chapter 3, private developers can select a contractor based on lowest bid, qualifications, reputation, and/or past experiences. They are not legally required to select a contractor based purely on lowest bid and furthermore, are not required to solicit multiple bids for a project. However, for publicly funded construction projects public developers are often required by local or state ordinances to solicit multiple bids and select the lowest bid for the project or portions of the project that involve public funds. They can use qualifications to restrict the eligible contractors. In the case of a public project, one qualification

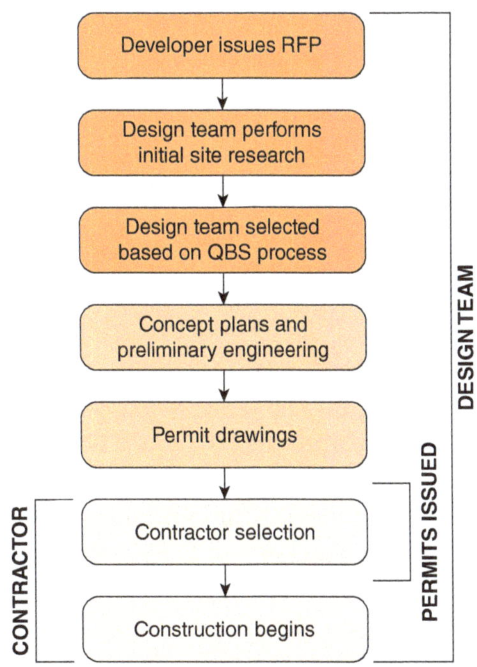

Figure 2.3 Design-bid-build process.

for a correctional facility may be previous experiences with correctional facility construction within a specified time and geographic area. Once qualified bidders are identified, the public developer is required to award the bid to the lowest bidder. In the case of a private residential project, a sanitary trunk sewer extension may be required, and the local jurisdiction may be willing to pay for a portion of the improvements. The local jurisdiction typically provides a list of qualified contractors and/or subcontractors for the private developer to procure at least three competitive bids and award the construction contract to the lowest bidder.

As design-bid-build is the traditional project delivery method, it is better understood by both the public and the construction industry as a whole and therefore, is often the default selection. The selection process for a contractor is an easier, more objective process as compared to other project delivery methods. For public projects, the local ordinances may require a low-bid selection process, which is more easily facilitated through design-bid-build.

From the design team's perspective, design-bid-build requires less effort prior to design contract execution as compared to design-build, which is when the developer has one contract for design and construction, selecting the design and construction team based on preliminary design and estimated cost. For this reason, a design-bid-build contract reduces risk for the design team. For a private development, the design team will prepare a proposal for design services for the developer's review whereas a proposal and interview may be required for a public development. In either case, the design team will do some initial site research and potentially conceptual plans but not provide as much in-depth analysis as is necessitated during the design-build selection process. It is extremely important for the design team to communicate design decisions and any special parameters to potential bidders and the selected contractor since the contractor is removed from the design process and therefore unaware of any planning decisions made early in the project that may impact construction.

For developers, design-bid-build provides additional time to finalize the requirements for the built project and allows for their input on the detailed design and specifications. The developer will typically hire the design team to prepare conceptual design documents to evaluate design alternatives prior to commencing final design. Both the concept plans and site plans are complete or nearly complete prior to bidding the project for construction. The developer may also perceive that this delivery method provides a lower upfront construction cost as the contractor is bidding on a complete or nearly complete set of site and building plans. Finally, it provides a checks and balance between the design team and the contractor as both have a contract with the developer but not with each other. The design team develops drawings and specifications to define a quality project meeting the developer's criteria whereas a contractor evaluates the design considering schedule and cost.

Some disadvantages of design-bid-build include time, potential cost, and quality. The design team typically finalizes the site and building plans including obtaining approval from the local jurisdiction prior to soliciting bids from contractors for construction. Therefore, it does not as easily accommodate an overlap of design and construction activities, which extends the project schedule as compared to other project delivery methods. The design team also does not have the contractor's experience with constructability and value engineering opportunities during the design development, so any necessary or developer-preferred design revisions in light of a contractor constructability review or value engineering proposal necessitate additional design fees. If these activities necessitate design revisions during construction, the contractor will request a change order for additional materials and labor associated with the revision. If the developer selects the contractor based on lowest-bid, the contractor may use alternative materials or less preferred construction practices to meet the contract price or seek opportunities like design revisions to add change orders, increasing the construction contract amount. At the developer's request, the design team reviews contractor bids and requests clarifications to help the developer make an informed decision.

2.2.2. Design-Build

With the design-build (DB) delivery method, a developer has two contracts for design and

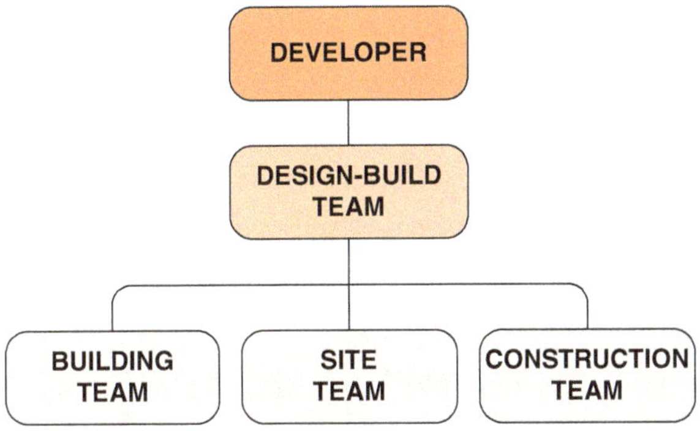

Figure 2.4 With a design-build delivery method, the developer has one contract for design and construction with the design-build team. The design team and contractor are part of the same firm, work together as a consortium, or contract between themselves with subconsultant agreements.

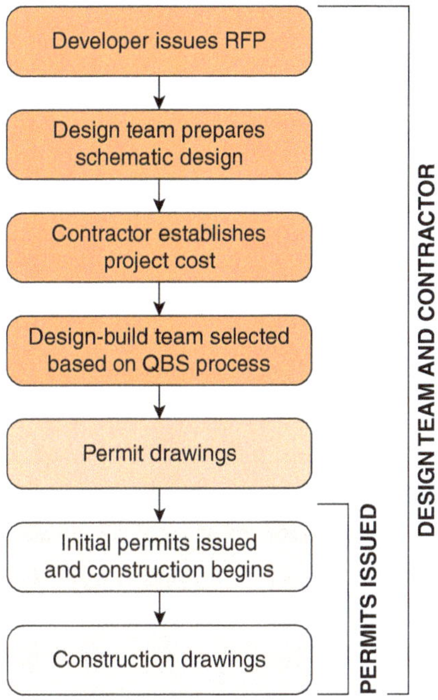

Figure 2.5 Design-build process.

construction—one with a criteria consultant to develop the project specifications and schematic documents and the other with the design build team, as shown in Fig. 2.4. The contract may be with a firm that does both design and construction or may be a contractor and civil engineer who have formed a design-build team. The typical scenario is where the contractor is the design-builder, holding the prime contract with the developer, and the civil engineer is a design subconsultant to the contractor.

The contract selection process is initiated through a request for proposals (RFP) process where the level of detail provided must be adequate for the civil engineer to begin the schematic design of the project (see Fig. 2.5). Unlike a design-bid-build project where the plans and specifications define the project scope, the specific performance criteria within the RFP define the scope for a design-build project. Developers may prepare this performance criteria or contract with a criteria consultant to develop project specifications and schematic documents prior to issuing the RFP for the design-build team. The developer selects the design-build team based on a qualifications-based selection (QBS) process considering experience and technical expertise and then negotiates the price. Qualifications may include previous design-build projects, projects of similar use, or projects of similar scale. The selection of the winning team may require multiple review iterations and interviews. Based on the RFP, interested design-build teams will prepare a schematic (10–30%) design for the developer's consideration. As the design effort to win the project far exceeds the requirements of other delivery methods, a developer may provide a stipend to the unselected teams to cover some of their design costs, but otherwise the team will absorb this marketing pursuit cost.

A design-build contract can be lump sum, cost-plus, or cost-plus with a guaranteed maximum price (GMP). With a lump-sum contract, the design-build team provides one price to the developer to construct the entire project based on the prepared schematic plans and specifications. With a cost-plus contract, the developer reimburses the design-build team for any costs plus a predetermined fee, typically a percentage of the cost. A cost-plus with a GMP is similar except it sets a limit on the amount a developer will pay for the project. In this scenario, the design-build team is responsible for any costs exceeding the GMP, and if the project costs do not exceed the GMP, often times the contract will stipulate how savings will be distributed between the developer and design-build team.

A design-build delivery method can offer a reduced project schedule. By selecting the design team and contractor concurrently, early in the project development, construction may begin prior to the completion of the design documents. Typically, once the local jurisdiction has approved and permitted the site plans, the contractor may begin clearing and grading the site while the architect finalizes the building plans. Similarly, the construction of the building can begin once the local jurisdiction issues the building permit, and an interior designer finalizes the interior finishes while construction is ongoing. A developer may seek a faster project delivery method due to funding source deadlines or based on revenue gained by an earlier project completion.

From the developer's perspective, a design-build delivery method offers an earlier, confirmed project cost and reduced conflict between the design team and contractor. The project cost is set prior to contract ratification when using a cost-plus with a GMP or a lump-sum contract. Combining the design team and contractor under one contract can eliminate conflicts inherent with a design-bid-build delivery method where there is no contractual relationship between the two parties. The design team and contractor are working together to meet the specifications from the RFP and to meet the contract terms with the developer. Conversely, the developer may lose some of the checks and balances present in the design-bid-build delivery method when both parties are contracting directly with the developer. Similar to design-bid-build, contractors will prioritize cost over quality to ensure they achieve the GMP. The developer also may not have as much influence over the specific design details to meet the RFP criteria as compared to a design-bid-build project. Changes to the performance criteria may result in a change order and therefore additional costs beyond the GMP. Other benefits of design-build for the developer include:

- Reviewing multiple design solutions before awarding contract
- Integrating design solutions from unselected design build teams into final design
- Shifting risk of constructability and accuracy of the design to the design-build team
- Selecting the design team and contractor based on qualifications in addition to cost
- Publicly releasing only a RFP rather than final construction drawings, thus alleviating security concerns

The design team benefits from the involvement of the contractor throughout the design process, offering constructability reviews and providing input on cost throughout the design. This collaboration continues into construction as both the design team and contractor work together to minimize risk rather than operating as two separate entities looking out for their own interests. However, the initial risk and effort required to acquire a design-build contract is greater for the design team as compared to design-bid-build. Prior to awarding the contract, the civil engineer completes at least a schematic design, which is at least a 10–30% design. This requires a site survey, site research, and detailed design in addition to the project scope and fee to respond to the RFP. If provided, the developer's stipend will not reimburse all of the labor cost exhausted to respond to the RFP or the design team's typical profit margins. Additionally, as the design team works for the contractor instead of the developer, there may be additional pressures to reduce design fees to avoid exceeding the overall project budget. The design team should carefully review the contract terms to understand who is responsible for the ultimate performance of the project and negotiate revisions to the contract as needed to reduce liability for construction issues. Design teams pursue design-build contracts in spite of the risks as some industries use it exclusively; however, due to the cost and inherent risk of pursuing these projects, a civil engineer must be selective in these pursuits.

2.2.3. Construction Manager at Risk

With the construction manager at risk (CM at risk or CMAR) delivery method, a developer will have a contract with the design team and two contracts with a contractor, one for preconstruction services and one for construction as shown in Fig. 2.6. The contractor's preconstruction services may include cost estimating, constructability reviews of the site

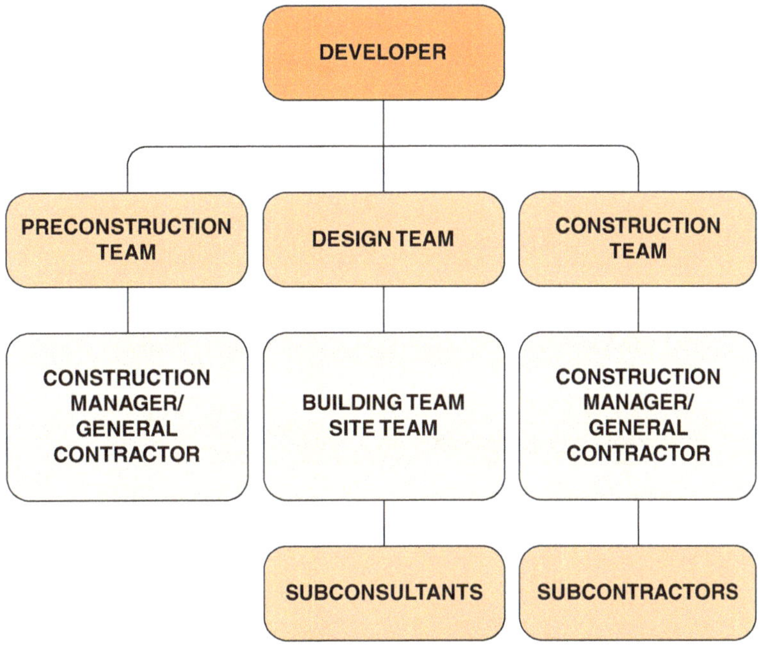

FIGURE 2.6 Construction manager at risk delivery method. The developer has one design contract and two construction contracts–one for preconstruction services and one for the actual construction. The construction manager, who serves as the general contractor during construction (CM/GC), is preferably the same company for preconstruction and construction but can be two different entities.

plans, and establishment of the guaranteed maximum price (GMP). The GMP sets the construction budget as the maximum amount the developer should expect to pay for a completed project, including specified allowances and contingencies, and is established at a mutually agreed-upon time during design development. As shown in Fig. 2.7, the GMP is typically set before site plans are complete and approved by the local jurisdiction. After accepting the GMP, the developer issues a second contract with a contractor for the construction services. Ideally, the same contractor who performed the preconstruction services will also perform the construction, but if the developer and contractor cannot agree on the GMP, the developer may select a different contractor for construction.

The CM at risk project delivery method is not the same as contracting with a construction manager (CM) to be onsite as the developer's representative throughout construction. In this case, the CM is a separate entity from the GC who has a construction contract with the developer. The contractor, not the CM, performs the construction activities. With CM at risk, the general contractor is the construction manager (CM/GC) and signs construction contracts with subcontractors.

CM at risk combines components of both design-bid-build and design-build into one delivery method. CM at risk maintains the checks and balances of a design-bid-build delivery method by having the design team contracted separately from the contractor. However, it incorporates the contractor's knowledge of means and methods, materials, and pricing during the design phase similar to design-build. The contractor's involvement in preconstruction may reduce the risk of cost overrun or an extended construction schedule. Value engineering completed throughout the design phase with the contractor's input can identify design components or decisions that impact cost and/or schedule, and the design team can correct these items prior to bidding the construction contract. These contractor reviews may necessitate additional design revisions by the civil

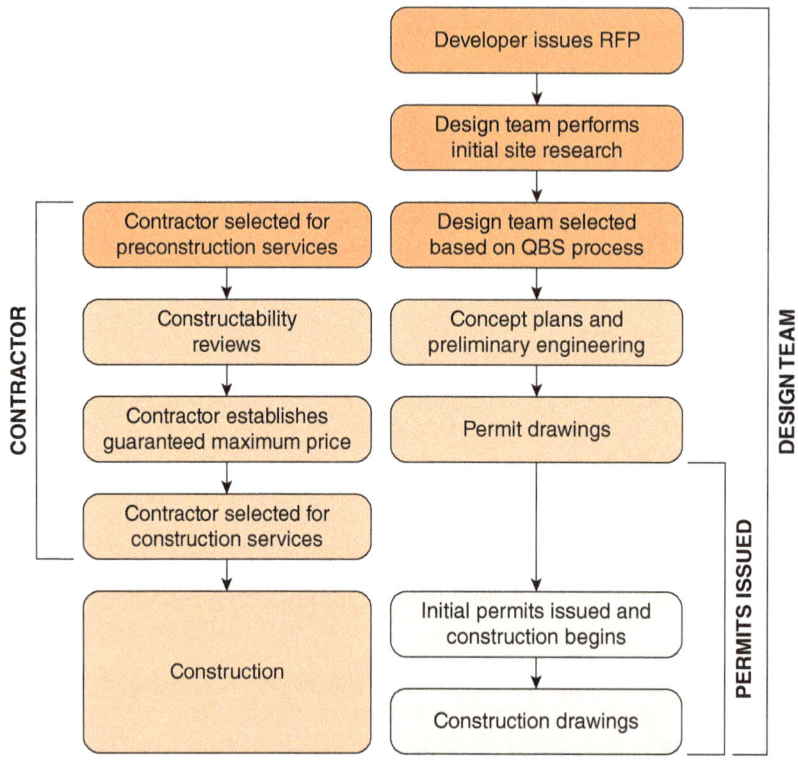

FIGURE 2.7 Construction manager at risk process.

engineer in addition to other revisions necessitated by the local jurisdiction's review. Additionally, the developer's risk is reduced as intermittent contractor reviews of the site plans prior to construction is intended to address constructability concerns and increase accuracy. As with design-build, construction may also start earlier than design-bid-build with appropriate phasing, since complete construction drawings may not be required to bid the construction contract.

In order to achieve the maximum benefit of this project delivery method, the design and construction team should remain consistent throughout the project duration. In this scenario, the contractor will be aware of design decisions and sensitive site constraints prior to construction. However, if there are unresolvable issues between the developer and contractor including an agreed-upon price for the construction, the developer has the option to change contractors prior to construction. The developer should also be aware that when the design is incomplete, and therefore subject to change, the contractor may inflate his GMP in order to protect his interests and account for any uncertainties in the contract.

CM at risk may not be appropriate for every project due to the additional upfront cost. For example, it may not be applicable to smaller, less complex projects that would not receive the full benefit of a contractor's involvement during the design phase.

2.2.4. Public-Private Partnership

For a public project, a government agency or local jurisdiction may use a public-private partnership (P3) to contract with a private entity to provide some portion of the project, including financing, design, construction, and/or operations and maintenance as shown in Fig. 2.8. The private entity is typically a project-specific joint venture (JV) or limited liability company (LLC) formed solely for the project. It consists of a combination of equity investors, contractors, and/or concessionaries with the design team typically under the direction of a design-builder who may or may not have equity interest in the project. The private entity shares a significant amount of financial, technical, or operational risk thereby allowing the public agency to maintain its public services for its citizens without

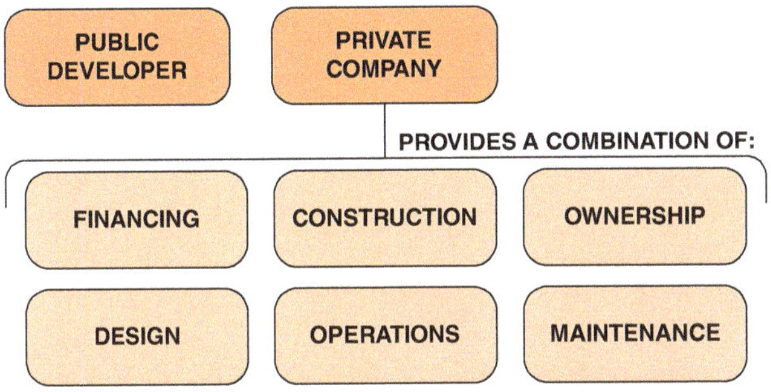

Figure 2.8 With a public-private partnership, a private company provides some combination of the public project's financing, design, construction, operations, maintenance, or ownership responsibilities and therefore a portion of the project risk.

assuming all of the initial project costs or risk. Three common types of P3 contracts include design-build-maintain, design-build-operate, and design-build-finance-operate-maintain.

A design-build-maintain (DBM) delivery method is similar to design-build with an added maintenance clause for a certain period included. The maintenance clause further guarantees the finished project, thereby reducing long-term risk for the public agency.

With a design-build-operate (DBO) delivery method, a public agency contracts with a private company to design, construct, and operate the facility. As compared to a design-bid-build contract, DBO promotes continuity through all phases of the project and also relieves the owner from responsibility for operations. The private company may recuperate its financial investment in the project through operational fees charged to facility users. The developer can also include a maintenance clause, which results in a design-build-operate-maintain (DBOM) contract. In either case, the jurisdiction remains the owner of the project following project completion.

In contrast with the DBM or DBO, the design-build-finance-operate-maintain (DBFOM) delivery method transfers financial responsibility for the project to a private entity while the jurisdiction remains the owner of the facility throughout the project. With DBFOM, the project is funded through a combination of a debt-leveraging revenue stream, bonds, or public grants. A debt-leveraged revenue stream is where private companies invest in a project based on the forecasted future revenue stream (such as tolls on roadways) that will repay the investment in addition to covering operations and maintenance cost. Public grants for P3 projects may include money or providing land for right-of-way.

Under DBM, DBO, and DBFOM, the public agency will remain the owner of the project. However, some P3 delivery methods such as build-operate-transfer and build-own-operate transfer ownership to a private company either in perpetuity or for a specified period.

For a public agency, a P3 delivery method shares risk with private companies and helps overcome any funding gaps for projects. By involving private investors, the capital required for the project can be raised in a shorter period than through public funding mechanisms. P3 projects also accommodate gaps in public staff's availability or capability of operating or maintaining a project.

CASE STUDY: CONSTRUCTION MANAGER AT RISK

ISSUE:

A contractor's right to rely on drawings and specifications furnished by owner, under implied warranty of design, through a construction manager at risk delivery method.

SUMMARY:

A1 Construction Management was the construction manager at risk (CMAR) on a public hospital project. First Electrical was a subcontractor to A1 Construction Management. According to First Electrical, errors and omissions in the design furnished by the public owner (and prepared by the design team) were a primary cause for labor increases of 49% for the electrical work. A1 Construction Management ultimately pursued recovery from the owner for the losses incurred by the subcontractor.

In traditional design-bid-build construction, the owner presents the contractor with a completed set of drawings and specifications, and impliedly warrants that design against defects. As stated nearly a century ago in United States v. Spearin, "if the contractor is bound to build according to plans and specifications prepared by the owner, the contractor will not be responsible for defects" in the design. In the CMAR project delivery method, however, the CMAR typically has responsibilities with respect to the design, while the architect/engineer (A/E) is still preparing it. During the design phase, a CMAR typically provides advice, consultation, and commentary regarding the design, and may influence the final design. In this case, the public owner argued that the CMAR's involvement during the design eliminated the CMAR's ability to benefit from an implied warranty of the design. The trial court agreed with the owner's position, finding that the CMAR method entails "material changes in the roles and responsibilities" of the parties, and that as a result the owner did not impliedly warrant the design to the CMAR.

DECISION:

The Massachusetts Supreme Judicial Court reversed the lower court decision, holding that the owner continues to impliedly warrant the design under CMAR. The court reasoned that despite CMAR consultation during the design, the owner (and the owner's A/E) ultimately control the design. Mere consultation should not make the CMAR a "guarantor" against all design defects.

Despite this general support for the implied warranty of design in the CMAR setting, the court did hold that the scope of the implied warranty is somewhat reduced in CMAR. In limiting the scope of the implied warranty, the court noted that even under the law applicable to implied warranty of design in a traditional setting, the warranty is not absolute: the contractor's reliance on the design must be given in good faith, essentially meaning that a contractor may not ignore obvious omissions or errors in the design. In a CMAR situation, the court held that there must be good faith reliance on the design and the contractor must act "reasonably" in light of its own "design responsibilities." According to the decision, the greater the CMAR's design responsibilities in the contract, the greater the burden on the CMAR to show that its reliance was in good faith and reasonable.

COMMENT:

A CM or GC is not a design professional and should not have genuine "design responsibilities." Most of the design-related duties under the A1 Construction Management CMAR contract are similar to those required of the contractor under standard general conditions such as EJCDC® C-700—for example, careful review of the design, comparison of the design to field conditions, and reporting of obvious errors or inconsistencies. The difference is that the CMAR begins its design-related tasks before the design is completed, at a point when the CMAR input can make a greater impact (and in some cases before a final construction price or GMP is established). In addition, the CMAR looks for the potential impact of the design on construction costs, and constructability. But it is important to recognize that a CMAR is not qualified to apply engineering expertise to a review of the design. A CMAR's review of the design is not a peer review. A CMAR contract should be explicit in emphasizing the two very different roles played by the A/E and the CMAR, and in fact the A1 Construction Management contract had some good clauses in that regard (for example, a statement that the CMAR shall make recommendations about the design relating to "construction feasibility, schedules, cost or quality...without, however, assuming the [A/E's] responsibility for design.")

The "limits" that the Massachusetts decision places on the scope of the implied warranty in CMAR appear to be reasonable and unremarkable. Essentially, if a CMAR's consultations regarding design result in exposure of a possible defect or similar issue, the issue should be confronted and addressed at that point, for the good of the project. If the CMAR raises an issue and the A/E and owner choose to ignore it, the implied warranty should apply.

Case study based on Coghlin Electrical Contractors, Inc. v. Gilbane Building Company. *Supreme Judicial Court of Massachusetts (2015).*

The information provided does not constitute legal advice. Consultation with an attorney is strongly advised.

2.3. CONCLUSION

Typical considerations for selecting a project delivery method include cost, schedule, project size, funding sources, and risk transfer. No single project delivery method is appropriate for every project and therefore is selected based on project constraints, full range of risk factors, and developer preferences. While design-bid-build is the traditional delivery method and most commonly used, newer delivery methods such as design-build and construction manager at risk address some shortcomings of deisgn-bid-build such as schedule and lack of collaboration between the design and construction teams. A challenge for a developer pursuing design-build or public-private partnership projects is unclear funding. Many public projects cannot commit to the full project funding until they have more detailed construction cost estimates in addition to political support.

CHAPTER 3

PRECONSTRUCTION ADMINISTRATION

3.1. INTRODUCTION

The process to procure a contractor to complete the construction of the project can vary from project to project. Public projects require the construction contract to be competitively bid through advertising, reviewing, and awarding the bid based on established procedures and laws. This prevents sole sourcing all projects to a select group of contractors while also maintaining a competitive market for equitable use of public funds.

On the other hand, private developers may choose to bid their project or directly engage a contractor. A private developer can sole source the construction contract by soliciting a price from just one contractor that they have previous experience with or prefer to work with. Private developers can also have a private bid process where they request bids from select contractors, again based on reputation or previous experiences, and select any of the received bids to negotiate the construction contract. They may publicly advertise the bid and any qualified contractor may submit a bid. Again, they may select from any of the received bids and are not bound to the lowest bid.

Additionally, private or public developers may choose to interview contractors as part of the selection process. During this interview, contractors may suggest design alternatives they believe can save cost or time, and the developer can ask questions specific to the contractor's team. This is more common with design-build projects which consider qualitative merits such as past performance, references, project teams with previous experience working together, and key staff experience and technical skills in addition to cost.

Prior to bidding the project, the developer in coordination with the design team must compile the construction documents, which include all the documents used for bidding and ultimately construction. These documents are broken into three categories—the project manual, bidding documents, and contract documents—as shown in Fig. 3.1.

Prior to issuing the project manual and advertising the bid, the civil engineer should work with the developer to determine the following:

- Minimum bid period—time between the bid advertisement and bid opening
- Mandatory versus nonmandatory pre-bid meeting—typically occurs in first third of the bid period
- Deadline for bid questions
- Deadline to issue addenda
- Bid opening date

Once the project manual and construction drawings are complete, the developer publishes an invitation to bid or bid advertisement online, in trade journals, and/or in the newspaper. Alternatively, a developer may decide to prequalify contractors

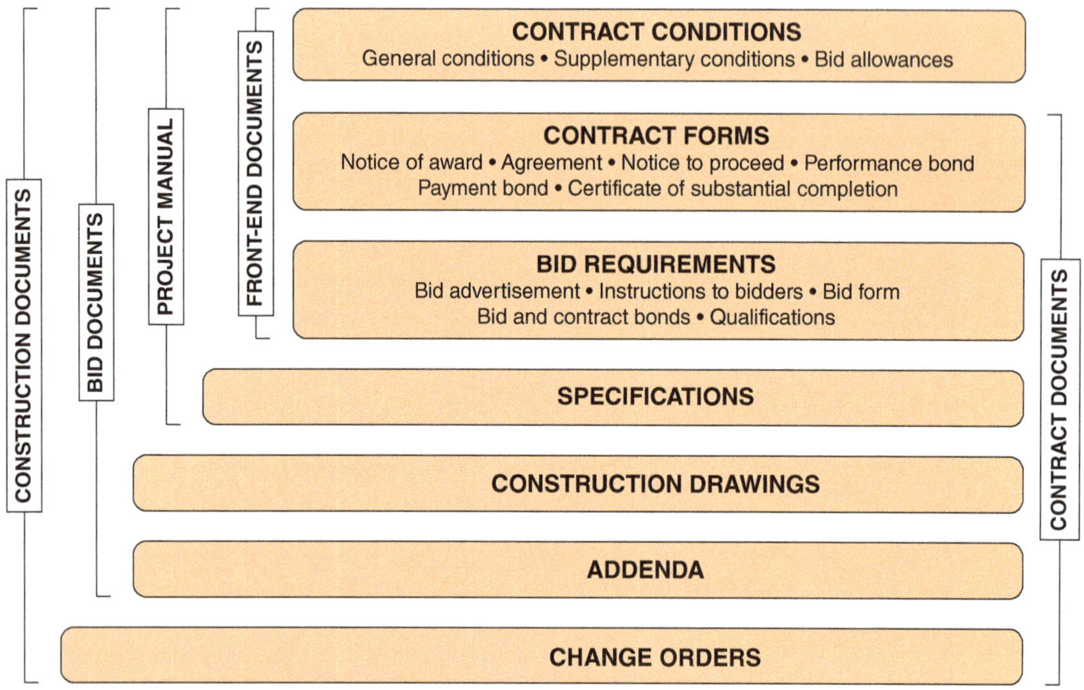

Figure 3.1 Construction documents.

prior to bidding the project, in which case, only the prequalified contractors will receive the invitation to bid.

Contractors have until the bid opening to review the construction drawings and project manual and to ultimately submit their bid. A developer may hold a pre-bid meeting for all interested bidders to provide an overview of the project and solicit any questions. Attendance at pre-bid meetings by prospective bidders may be mandatory or optional.

After the bid opening and prior to awarding the contract, each bid will be reviewed for completeness and adherence to the requirements outlined in the project manual. A contractor will be selected based on the lowest bid or other criteria selected for design-build or privately owned projects. All permits required for land disturbance, utility tie-ins, and driveways, as well as bonds must be in place before commencing construction. A building permit is not required to start site work and can be acquired later after building plans are approved. A preconstruction meeting onsite with the local jurisdiction is also typically required. The bid process into construction is shown in Fig. 3.2.

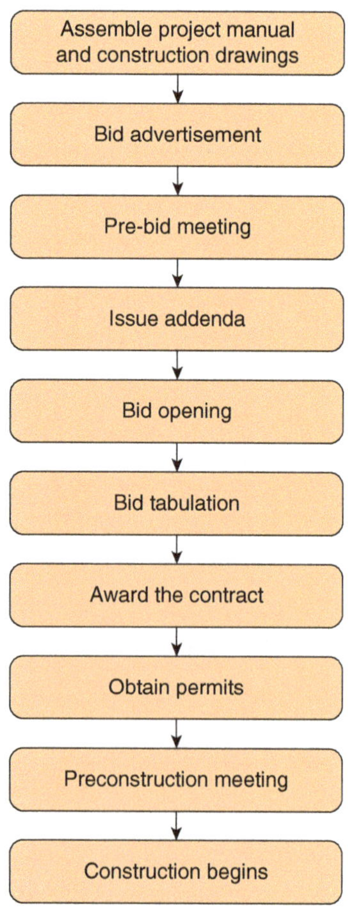

Figure 3.2 Typical process for a publicly bid project from design to construction commencement.

The following terms will be used throughout this chapter:

Responsive Bidder	On time, meeting all requirements of bid including bid and contract bonds, forms, and acknowledgment of any addenda
Responsible Bidder	Meets all contractor qualification requirements outlined in the bid documents and is capable of providing performance and payment bonds
Agreement	The construction contract between the developer and the contractor
Base Bid	The lump-sum price or combination of lump-sum and unit prices the contractor offers the developer to complete the scope of work defined in the bidding documents
Bid Alternate	An addition or variance to the scope of work outlined in the bidding documents for which the contractor should provide a separate, lump-sum price relative to the base bid
Guaranteed Maximum Price (GMP)	Agreed upon price between the developer and contractor for construction, determined prior to design completion, and only modified afterwards with a change by developer or designer
Lump Sum	A set, all-in-price established for a defined scope of work
Unit Price	A price established per material quantity and/or task without a set, all-in-price as established with lump sum
Cost Plus	A contract paid through unit prices for the line item materials required to complete the project with expenses paid separately, as incurred
Cost Plus Max	A cost-plus contract with a maximum price established based on the defined scope of work
Bonds	A surety provided by the contractor to the developer or local jurisdiction to protect against the contractor failing to meet the contract requirements or specifications as well as to protect the developer from payment claims from the contractor's subcontractors

3.2. PROJECT MANUAL

The project manual includes the bid documents detailing bidding requirements, contract forms, contract conditions, and the project specifications. Often, the bid requirements, contract forms, and contract conditions are referred to as the front-end documents. As previously shown in Fig. 3.1, bid requirements include the bid advertisement, instructions to bidders, and bid forms. The contract forms include the agreement, performance and payment bond, and certificates. Contract conditions are separated into general and supplementary conditions, and the specifications are divided into general requirements and technical specifications. The project manual may utilize a standard form such as the Engineers Joint Contract Documents Committee (EJCDC) or the American Institute of Architects (AIA). Alternatively, the project manual may be based on a developer-defined template, such as those typically required for state-funded projects or well-established commercial retailers.

3.2.1. Bid Advertisement

Public projects follow a prescriptive bid process where the project must be publicly advertised and typically the lowest responsive and responsible bidder is awarded the contract. State law or the local jurisdiction's code typically requires the bid advertisement be publicly advertised through government websites, the newspaper, online plan rooms, and/or construction trade journals prior to the bid opening, also known as the bid letting date. Each state's and local jurisdiction's requirements vary, but typically the bid advertisement must be released no sooner than 10 days or two weeks before the bid opening. Other names for the bid advertisement include notice to bidders and invitation to bid. Typical components of the bid advertisement include the following:

- Project Developer or Responsible Agency
- Project Name
- Bid Opening Date, Time, and Location
- Pre-bid Meeting Location, Day, and Time (if applicable)
- Availability of Plans and Specifications
- Required Qualifications
- Bid Requirements

An example bid advertisement is shown in Fig. 3.3.

3.2.2. Instructions to Bidders

The instructions to bidders document contains information on the bid forms, requirements, and procedures. It refers to other construction documents such as the supplementary conditions, agreement, and bid advertisement where project-specific information can be reviewed. The table of contents from the EJCDC standard forms is included in Fig. 3.4.

3.2.3. Bid Form

The bid form repeats most of the information provided in the bid advertisement in addition to outlining bidder acknowledgments, basis of bid, and any required attachments. Based on the EJCDC Bid Form document, the following sections are included in a bid form:

Bid Recipient	provides developer's information and acknowledgment that bidder will enter into an agreement with the developer if selected
Bidder's Acknowledgments	bidder agrees to terms within the instructions to bidders, and specifies that bid is valid for up to 60 days
Bidder's Representations	statement that the bidder has reviewed available bid documents including addenda and construction drawings as well as visited the site; also bidder confirms identifying any issues with the site plans and agrees with the civil engineer's proposed solution
Bidder's Certification	certifies the bidder prepared bid in good faith without collusion with other bidders or engaging in any corrupt practices
Basis of Bid	conditions showing that the bidders provided unit price and/or lump-sum prices for work; if any components of the bid are to be completed as unit price, they will submit a table similar to Fig. 3.5. The unit costs provided are typically the installed cost (materials and labor) unless otherwise specified, even for a lump-sum bid, a developer may request a unit price breakdown for major elements as a way to compare bids and also to use to evaluate any future potential change orders (PCO). However, it is rare that a contractor will provide a full unit price breakdown of all items even when provided with a standard form in the project manual due to competitive advantage and flexibility on pricing a PCO during construction
Time of Completion	the duration of construction and statement that the bidder agrees to liquidated damages as outlined in the agreement
Attachments to This Bid	documents required in addition to bid form; examples include bid security and qualifications statement
Bid Submittal	bidder's signature and contact information

INVITATION TO BID

Sealed bids for the construction of the Main Street Recreation Center project will be received by Washington County at the Parks and Recreation Department office, 123 Main Street, Washington County, until 2:00 PM local time on April 1, 2019, at which time the Bids received will be publicly opened and read. The project consists of constructing an approximate 87,000 sf state-of-the-art recreation facility including a gymnasium, fitness areas, indoor pool, locker room, and multipurpose rooms as well as associated site improvements including parking, utilities, and stormwater management.

A mandatory pre-bid meeting will be held at the Washington County Parks and Recreation Department office at 10:00 AM on March 15, 2019. The cutoff date for pre-bid questions will be Friday, March 22, 2019, at 5 PM.

A contractor license and registration is required in accordance with the state code. See the instructions to bidders for additional contractor qualification requirements.

The issuing office for the bidding documents is Washington County Parks and Recreation Department, 123 Main Street, Washington County. Prospective bidders may examine the bidding documents at the issuing office Monday through Friday between the hours of 9:00am and 4:00pm or purchase a printed copy of the bidding documents during these hours for $100 (checks made payable to Washington County Treasurer). Neither the Owner nor Engineer will be responsible for full or partial sets of bidding documents, including addenda if any, obtained from sources other than the issuing office.

Bids will be received for a single prime contract. Bids shall be lump sum and unit price basis as indicated in the Bid Form. Bid Bond and Contract Security will be required in accordance with the Instructions to Bidders. Additional procedures for submitting a bid, claiming an error, withdrawal of bids and other pertinent information are contained in the Instructions to Bidders, which is part of the Invitation for Bids. Withdrawal due to error in bid will be permitted in accordance with Section 9 of the Instructions to Bidders and state code. The Owner reserves the right to reject any or all bids.

END OF NOTICE OF INVITATION FOR BIDS

FIGURE 3.3 Example bid advertisement.

3.2.4. Bid and Contract Bonds

Bid bonds may be required by the developer from contractors as a surety that they will accept the project if selected as the lowest bidder. Bid bonds cover the difference between the contractor's bid amount and the next lowest bid, up to a set percentage (typically 5%) of the bid. As with other bonds, the contractor will pay a nominal fee to the bonding company to post the bid bond. A bid bond reduces the risk to the developer of the lowest bidder declining a project after realizing the discrepancy in the low-bid price versus the other bidders.

A contract bond guarantees that the contractors will provide the required performance bonds, payment bonds, and insurance policies if they are the successful bidder and therefore awarded the construction contract.

3.2.5. Qualifications

If developers do not have a prequalification process, they may include qualifications within the bidding requirements, typically specified within the instructions to bidders and the bid form. In this case, any contractor who submits a bid but does not meet the qualifications may be deemed nonresponsible, and therefore, the contract will be awarded to the lowest, responsible bidder. While the contract will still be awarded based on the lowest responsive and responsible bid, using a prequalification process or qualifications ensures the developer that the lowest bid contractor has the experience, authority, workforce, and equipment to complete the project. A work experience qualification may include a specified number of projects in this discipline or type of project completed within a specified time. An example qualification is requiring the contractor to have completed three higher education projects in the last two years in order to bid on a proposed academic building on a college campus. Another qualification may be that the contractors have the appropriate security clearance for a government project or that they have the licenses, such as a business license, to complete work in the jurisdiction. Finally, public developers may require a subcontracting plan that includes a disadvantaged business enterprise (DBE), which may also be referred to as a minority, veteran, or woman-owned company. The qualifications may stipulate

INSTRUCTIONS TO BIDDERS

TABLE OF CONTENTS

EJCDC® C-200, Suggested Instructions to Bidders for Construction Contracts.
 Page i

FIGURE 3.4 Suggested instruction to bidders for construction contracts—Table of Contents. Reprinted by permission of the Engineers Joint Contract Documents Committee (EJCDC). All rights reserved. For additional information, please visit www.ejcdc.org. (EJCDC C-200, 2013, i).

Item No.	Description	Unit	Estimated Quantity	Bid Unit Price	Bid Price
18	48" RCP Culvert, O Ring, CL III	LF	130		
19	6'x3' Precast Box Culvert	LF	117		
20	Headwall	EA	2		
21	2'-6" Curb and Gutter	LF	5770		
22	5' Sidewalk	LF	5230		

Figure 3.5 Example bid form unit price table. Reprinted by permission of the Engineers Joint Contract Documents Committee (EJCDC). All rights reserved. For additional information, please visit www.ejcdc.org. (EJCDC C-410 2013, 17).

a minimum utilization goal (typically specified as a percentage of the total cost of the contract) for a certain portion of the work completed by DBE subcontractors. The goal may be either required or set as an aspirational goal with good faith efforts to meet. Prime contractors bidding the project are responsible for identifying how they will meet the goal in their bid submission. Depending on the scope and schedule of the project, it may be important for a developer to know the workforce and equipment the contractors have to ensure they are capable of completing the project within the agreed-upon schedule. The developer may request a resume for the superintendent as well as qualifications for subcontractors to confirm their experience with similar size and scope projects.

3.2.6. Contract Forms

The project manual contains copies of the contract forms including the notice of award, agreement, notice to proceed, bond forms, and certificate of substantial completion. The notice of award informs the bidder that they are selected to move forward into contract negotiations and also stipulates that they have a certain time period to respond to the notice of award to avoid surrendering their bid bond.

The agreement is the contract for construction. It outlines project information similar to what was presented in the bid advertisement and bid forms. The agreement also outlines the contract milestones including an allowable time frame to reach substantial completion and any intermediate deadlines, liquidated damages, and incentives for early finishes. EJCDC defines substantial completion as "the time at which the work… has progressed to the point where, in the opinion of [the] Engineer, the work… is sufficiently complete, in accordance with the Contract Documents, so that the Work… can be utilized for the purposes for which it is intended" (C-700 2013, 4). Liquidated damages outline the penalty to the contractor owed to the developer when construction extends beyond the time allowance specified in the agreement plus any time extensions approved by the developer. Liquidated damages are typically a cost per day penalty for each day beyond the contract time that the contractor pays the developer. Some developers may offer an early-finish incentive to the contractor for each day the project is finished before the agreed-upon completion date. Similar to liquidated damages, this is typically offered as a financial reward per day in the agreement. This is particularly advantageous for projects with substantial public impact during construction such as a roadway construction project.

The agreement will also include the lump sum bid price of the selected contractor as well as any unit price information included on the bid form. It will outline payment procedures including retainage and final payment. Retainage is a small percentage (typically around 5%) of the requested payment by the contractor that the developer withholds until substantial completion, and the agreement sets the retainage percentage. The agreement outlines the contractor's representation such as the contractor has reviewed the contract documents, performed a site visit, and coordinated any errors or discrepancies with the civil engineer that were discovered in the plans when the contract was awarded. This is similar to the bidder representations from the bid form.

The notice to proceed specifies the start date for commencing construction and is issued to the contractor once the executed agreement is received by the developer. The notice to proceed may initiate a submittal or shop drawing review period in advance of the construction start date. This occurs when the onsite construction window is constrained and completing all submittal requirements in advance aids in meeting the overall construction schedule.

The project manual will also include copies of the bond forms. Bonds, or sureties, are a payment provided by the contractor to the developer or local jurisdiction to protect against the contractor failing to meet the contract requirements or specifications. They also protect the developer from payment claims from the contractor's subcontractors. Bonds are typically issued by insurance companies, and any claims against the bond will impact the contractor's credit rating and ability to acquire bonds for future projects. The cost of the bond is typically a small percentage of the bond amount, which will vary based on the perceived riskiness of the bond.

In addition to the bid and contract bonds, performance bonds and payment bonds are also common in construction. Performance bonds are a form of collateral payment posted by the contractors for the total value of the bid. If they are unable to finish the project, the developer can use the performance bond to hire another contractor to finish the work. Payment bonds are required of the contractor to ensure payment of any subcontractors. If the general contractor fails to pay the subcontractors, the subcontractors are paid from this payment bond rather than seeking payment from the developer.

Other contract forms that may be included in the project manual include the payment application, request for information (RFI) form, potential change order (PCO) form, and certificate of substantial completion. These are discussed in further detail in Chapter 5.

3.2.7. Contract Conditions

Following the agreement and contract documents, the contract conditions are broken into general and supplementary conditions. The general conditions are typically prepared by AIA or EJCDC, and the supplementary conditions modify the general conditions for the specific project. No project-specific modifications are made to the general conditions as this is all noted through the supplementary conditions. Based on EJCDC (2013), the contents of the general conditions include:

1. Definitions and Terminology
2. Preliminary Matters
3. Documents: Intent, Requirements, Reuse
4. Commencement and Progress of Work
5. Availability of Lands; Subsurface and Physical Conditions; Hazardous Environmental Conditions
6. Bonds and Insurance
7. Contractor's Responsibilities
8. Other Work at the Site
9. Owner's Responsibilities
10. Engineer's Status during Construction
11. Amending the Contract Documents; Changes in the Work
12. Claims
13. Cost of the Work; Allowances; Unit Price Work
14. Tests and Inspections; Correction, Removal or Acceptance of Defective Work
15. Payments to Contractor; Set-Offs; Completion; Correction Period
16. Suspension of Work and Termination
17. Final Resolution of Disputes
18. Miscellaneous

The supplementary conditions reference the same article and section number as the general conditions and supersede any articles in the general conditions. Supplementary conditions modify the general conditions by deleting nonapplicable sections, editing content, or adding additional information. For example, General Condition *5.06—Hazardous Environmental Conditions at Site* states:

> A. *Reports and Drawings:* The Supplementary Conditions identify:
>
> (1) Those reports and drawings known to Owner relating to Hazardous Environmental Conditions that have been identified at or adjacent to the site; and
>
> (2) Technical Data contained in such reports and drawings. (EJCDC C-700 2013, 17)

The associated Supplementary Condition 5.06 states:

> Delete Paragraphs 5.06A in its entirety and insert the following:
>
> A. No reports or drawings related to Hazardous Environmental Conditions at the Site are known to the Owner. (EJCDC C-800 2013, 1)

In this example, the general conditions specifically direct the contractor to review the supplementary conditions for project-specific information. However, supplementary conditions are not always required.

If permits, reports, or other regulations are provided, the project manual should include a disclaimer stating these documents are provided for the bidders' information only, and they rely on this information at their own risk. The bidder should conduct any additional evaluation necessary to complete their bid. Examples of provided documents or permits include geotechnical reports and wetlands permits.

Bid Allowance. An allowance is a common method of assigning a set cost or cost equation to an unknown scope of work. It reduces ambiguity and strive to normalize the bids. In many contracts, allowances are used for items such as rock or unsuitable soils. Even with a thorough geotechnical evaluation, it is impossible to determine the exact quality and quantity of subsurface material that will be encountered during construction. For this reason, the bid documents may include an estimated quantity of a given item, such as 10,000 CY of unsuitable soil material, and the contractors provide a unit price for handling each cubic yard of unsuitable material in their bid. During construction, the contractor will be paid based on how much material is encountered. The allowance helps the developer determine a budget for the item during the bid phase; however, the contractor will be paid for the actual quantity of material encountered, even if it exceeds the estimated quantity used in determining the allowance. If the quantity of material encountered does not exceed the allowance, the contractor will only be paid for the actual quantity handled. The measurement of the allowance item is typically defined in the specifications, such as measuring unsuitable soils in situ, and unsuitable material may require an inspection and daily reporting by the developer's agent to the developer.

A developer may also include an allowance amount for services provided by another party such as temporary power supplied by the power company. By establishing a set price that all contractors include in their bid, the developer eliminates variability in the bids from inexperience, under scoping the work, or manipulating line items to achieve the low bid. As with an allowance based on a quantity, the developer will pay the actual cost of the temporary power, regardless of whether it is above or below the allowance amount set in the bid.

The instructions to bidders will detail how any allowances will be incorporated into the bid price. Allowances are specified within the general conditions and supplementary conditions as applicable.

3.2.8. Specifications

Technical specifications detail the "… written requirements for materials, equipment, systems, standards, and workmanship as applied to the Work and certain administrative requirements and procedural matters applicable to the Work (EJCDC, 2013)." Similar to the other construction documents, the specifications can be based on an industry standard such as AIA's MasterSpec. Specifications can also be developer-prepared or prepared by the design team based on similar clients or past projects such as higher education projects. Regardless of the origin of the specifications, it is important for the civil engineer to be familiar with the specifications as they supersede the construction drawings if there are discrepancies between the two. The specifications for the site work can be stand-alone or part of the overall project specifications, which include multiple disciplines. In the case where multiple disciplines are involved, the specifications should be cross-referenced between the site work and building requirements. As an example, the civil engineering design typically terminates five feet from the building, and the plumbing engineer is responsible for the plumbing design within five feet of the building. Both the civil engineer and plumbing engineer will have specifications on utility installation and materials, so the specifications should be reviewed and updated to eliminate any conflicting information for the installation process or material selection. Preferably, one discipline will defer to the materials, fittings, etc. specified by the other discipline rather than cross-listing the same information. This is particularly important as the site work subcontractor and building plumber will likely not be provided a complete copy of the specifications by the general contractor for bidding but rather just the specifications relevant to their work. Coordinating specification sections across disciplines mitigates conflicting information and potentially bid errors.

Each technical specification typically includes sections on general requirements, product or material selection, execution, and sometimes, approval requirements. The general requirements outline the parameters used to develop the specification, typically referring to industry standards such as ASTM. As the specifications should meet or exceed the local jurisdiction and state requirements, the general requirements may list references to any applicable local or state design standards. The product or material selection section will detail the required materials and/or manufacturers, cross-referencing any industry standards. For example, a waterline specification may state ductile iron pipe is required. The execution section will outline any applicable handling or installation requirements but is not intended to fully detail the contractor's means and methods. For example, the sanitary sewer manhole specification may require an epoxy coating (as detailed in the material specification) on the inside of the manhole. Similarly, any specification for using PVC pipe will outline storing the pipe such that it is shielded from sunlight. Lastly, the approval section, if included, will provide additional details on the submittal review and/or testing during the final inspections and punch list. Example specification sections as well as the entire table of contents for AIA's MasterSpec are included in Appendix B.

One particularly important specification section for site work is earthwork, which will indicate whether the site is classified or unclassified. In either case, the developer should provide the contractor with a subsurface investigation report prepared by a geotechnical engineer. However, the soil borings in this report represent only a sample of the site conditions. With an unclassified site, the contractor accepts any subsurface condition encountered during construction between the existing grade and proposed subgrade elevations shown on the site plans regardless of the amount of unsuitable soils or rock discovered. Additional undercut required beyond the subgrade elevation is considered separately, and the bid documents should indicate the depth of undercut the contractor should assume for the site based on the geotechnical report. As the contractor is accepting the unknown risk of the soils, an unclassified bid will be more expensive. A project bid with classified earthwork will define soil and rock excavation to be included in the base bid as well as provide an allowance or solicit a unit price for any unclassified earthwork the contractor encounters. While the initial bid will be lower with classified sites, potential change orders for unforeseen site conditions could increase the project cost above bidding the site as unclassified. The developer bids the site as classified with the expectation that the geotechnical report and site plans accurately represent site conditions encountered during construction.

3.3. BIDDING DOCUMENTS

The bidding documents include the project manual in addition to the construction drawings and addenda.

3.3.1. Construction Drawings

Construction drawings include any modifications or additions made to the site plan following site plan approval from the local jurisdiction (also known as approved site plans, permit drawings, or permit set). Permit drawings meet the local ordinances, whereas construction drawings may expand upon permit drawings to include additional, developer-requested details for the final site design. Developer-requested revisions or details may include enhanced landscaping plans or value engineering decisions that go beyond the requirements for site plan approval from the local jurisdiction. Depending on the timing of site plan reviews versus building plan review and approval, coordination with other disciplines including the architect and plumbing engineer may necessitate revisions after site plans are approved, and therefore, these revisions are incorporated into the construction drawings.

The construction drawings may also include bid alternates, which are either design alternatives for the same component or add-ons for the project. Examples of bid alternates may be material selection (gravel, asphalt, or concrete) for a multiuse path or adding a second back-up generator. By including bid alternates in the construction drawings, the developer receives an initial price from the contractor without including it in the base bid. Once developers have pricing on the bid alternates, they can select which bid alternates, if any, to include in the contract documents, thus avoiding implementing the changes after the executed contract. This is often done when the project budget for construction is very tight. If the add alternate is within budget, then developers can include it in the contract. Otherwise, they can elect not to do the alternate to stay within budget without having to negotiate or rebid the project with a reduced scope.

The construction drawings will also include any plan revisions necessitated by the addenda. Once the contract is executed, the construction drawings will become the contract drawings, which will include any plan revisions necessitated by a RFI or any developer requests. Revisions to the contract drawings are discussed further in Chapter 5.

Alternatively, the developer may choose to bid the project prior to receiving approved permit drawings from the local jurisdiction due to schedule constraints. In this case, the permit drawings are subject to revisions as a result of reviewer comments which could occur after awarding the bid. Any revisions to the permit drawings after the advertisement for bid will necessitate an addendum prior to bid opening or a change order after the bid award.

3.3.2. Addenda

Addenda are any changes to the bid documents after the bid package has been released to potential bidders. Examples of addenda revisions include changes to the project conditions or revisions to the construction drawings. Addenda most commonly address questions posed by the bidders through a RFI. Any addenda should be provided to all bidders with sufficient time to incorporate changes into their bid, which may require adjusting the bid opening date. After an addendum is issued, the civil engineer should not answer any additional questions unless it is deemed significant and/or will result in a high-value change to the bids. Where necessary, an additional addendum may be required as well as extending the bid period to allow all bidders the opportunity to review the addendum and update their bids.

3.4. PRE-BID MEETING

A pre-bid meeting is a meeting with the design team, developer, and potential bidders to review the project requirements. The bid advertisement will indicate if the meeting is mandatory. The pre-bid meeting is held after the bid advertisement but before the bid opening date. A pre-bid meeting will be led by the developer, the architect, or civil engineer, and will typically begin with an overview of the project and bid requirements. Members of the design team including the civil engineer may speak to important design considerations of the project as well. Next, the meeting may be opened up to solicit any questions from bidders, and the civil engineer should only answer questions by referencing bidders to the appropriate sections of the project manual, construction drawings, or specifications. Any other questions that require an interpretation or revision to the design should be submitted through a RFI and answered through an addendum to ensure all bidders receive the same information. Lastly, the pre-bid meeting is adjourned with a site visit, which depending on the project may be mandatory or optional for all bidders.

A sample pre-bid meeting agenda is included in Fig. 3.6.

3.5. BID WITHDRAWAL

Contractors can withdraw or modify their bid prior to the bid opening without forfeiting their bid bond. If the contractor wants to modify their bid, the instructions to bidders may specify the original bid must be withdrawn, and a new bid submitted prior to the bid opening date. Any withdrawn bids are returned to the contractor unopened.

For most projects, once the bids are open, contractors will surrender their bid bond if they choose to withdraw their bid unless otherwise specified in the instructions to bidders. It is uncommon for contractors to do this as any bonds pulled will decrease their bond rating, and therefore, affect their ability to post future bonds. The developer may allow for a short time period following the bid opening where contractors may withdraw their bid due to a substantial mistake in the bid preparation without losing their bid bond. This time period is specified in the instructions to bidders and is subject to the developer's approval of the mistake. In this case, the contractor may be disqualified from future bidding on the project if the developer chooses to rebid. For other projects, a developer may allow modifications to a bid after bid opening if the contractor notes a mistake such as a math error. Again, the developer's approval of the bid withdrawal is required to protect against contractors fabricating an excuse to withdraw their bid if they are the low bid with a large discrepancy with the other bids.

MANDATORY PRE-BID MEETING AGENDA

Main Street Recreation Center

March 15, 2019, 10:00 AM

123 Main Street

Washington County

1. INTRODUCTIONS
   - County Officials
   - Developer's Representative
   - Design Team
2. SIGN-IN SHEET
3. BID ADDENDUM
   - All questions are to be submitted in writing on or before Friday, March 22, 2019 at 5 PM.
   - Addenda, if any, shall be released Tuesday, March 26, 2019 by 5 PM.
   - Acknowledge receipt of Addenda and return with proposal.
4. BID OPENING
   - Sealed Bids for the construction of the Main Street Recreation Center project will be received by Washington County at the Parks and Recreation Department office, 123 Main Street, Washington County, until 2:00pm local time on April 1, 2019, at which time the Bids received will be publicly opened and read.
5. NOTICE OF AWARD
   - Bids remain subject to acceptance for 60 days.
6. NOTICE TO PROCEED
   - Notice to proceed can be issued when agreement is signed, but construction will not commence before May 6, 2019.
7. CONTRACT TIME
   - Work may not begin prior to May 6, 2019.
   - The work will be substantially completed on or before February 26, 2021.
8. REQUIRED BID ATTACHMENTS
   - Article 7 of Bid Form
9. BONDS AND INSURANCE
   - Refer to Supplementary Conditions for Workers' Compensation and Liability Insurance requirements
   - Performance Bond
   - Payment Bond
10. BID FORM
    - Total Bid Price should be included in Article 5 – Basis of Bid.
11. MAIN STREET RECREATION CENTER CONSTRUCTION REQUIREMENTS
12. BIDDER QUESTIONS
13. SITE VISIT (OPTIONAL)

FIGURE 3.6 Example pre-bid meeting agenda.

3.6. BID TABULATION AND REVIEW

At the bid opening, the developer reviews the received bids and determines the apparent low bidder. The bid opening may be privately conducted or open to all contractors who submitted bids. After the bid opening, the civil engineer may assist the developer in tabulating the bids. The bid tabulation will consist of comparing the total bid price in addition to any unit price breakdown provided by the contractor. The developer may also compare the bid tabulation to the latest engineer's estimate.

The bid forms should be reviewed for any exclusions in the contractor's bid, which may create inconsistencies among the bids. Exclusions are most often specified by the developer in the project manual, as any contractor perceived vagueness in the bidding documents should be addressed during the bid period and resolved with addenda as necessary. While uncommon, contractors may list additional exclusions based on their assumptions about the scope of work. The developer may choose not to allow any bidder added exclusions to the bid to ensure the basis of the bid is comparable across all bids. Any item that is clearly indicated in the drawings and/or specifications that the contractor or project manual does not designate as an exclusion is assumed by the developer to be included in the bid price. If contractors omit an item from their bid and also do not list it as an exclusion, then a potential change order is not warranted during construction for out-of-scope work. When permitted by the developer, it is at the contractor's discretion to list exclusions based on vagueness or assumptions on the scope of work.

No specifications may result in ambiguous bidding documents for the project. As an example, the construction drawings may indicate lawn areas on the landscape plan but because there is no associated specification on landscaping, the contractor may choose to seed lawn areas versus sod or exclude topsoil before seeding or planting. In this case, an exclusion would be sod and/or topsoil.

As contractors review the construction drawings, specifications, and other provided reports, they will incorporate the sequencing and means and methods necessary to complete the work in order to prepare their bid. Assumptions are typically made regarding the site conditions based on available information, often including the lowest-price solution in order to have a competitive bid. As an example, they may assume that they can reuse onsite material for fill in order to reduce costs versus importing offsite material. In this case, the exclusion is importing fill material.

Finally, based on their experience and interpretation of the construction drawings, contractors may exclude elements included in the construction drawings due to perceived illogical sequencing. For example, contractors may exclude constructing sidewalks for a residential subdivision project as the lots will not be brought to final grade as the houses will be constructed under a separate contract. In this case, they are delegating this work to the homebuilder rather than constructing a sidewalk which may have to be replaced once individual home lots are completed.

As exclusions are identified in the bid, the developer may request contractors to revise their bid to eliminate any or all of the exclusions. This will aid in comparing comparable bids to determine the true low bid.

During the bid tabulation, the developer will also check any references provided with the bid. Information revealed by the references on past projects will help to set the parameters for the relationship between the developer and contractor on the upcoming project.

3.7. AWARDING THE CONTRACT

As outlined in the instructions to bidders, the developer must award the contract within the acceptance period, which is a specified time frame after the bid opening. If the contract is not awarded within the acceptance period, contractors have the opportunity to withdraw their bid without surrendering their bid bond or adjusting their bid price. The developer will provide a written notice of award and agreement to the contractor. Oftentimes, the notice of award will stipulate the time frame to return the signed agreement to the developer.

Once the agreement is finalized, the developer will issue a notice to proceed, which will establish the construction schedule based on the start date provided in the notice to proceed and the time allowance until substantial completion stated in the contract forms. Typically, the notice to proceed either provides a specific start date or states the contractor shall begin work within a specified number of days from the date of the notice to proceed.

While atypical, unselected bidders can challenge the bid selection, though this is more likely to happen in a qualifications-based selection (QBS) rather than a low-bid contract. For design-build projects, points are assigned to potential contractors based on a technical proposal that includes a narrative addressing such factors as project approach, resource loaded schedule, and alternative technical concepts. The price proposal is negotiated separately. In this case, the contractor with the most points is selected, which may or may not be the low bid. The developer decides the weighting of each category considered, and since criteria such as technical merit or quality is subjective, it is very difficult for a contractor to contest the bid award. For a low-bid contract, contractors may contest a bid award if they believe the winning contractor included bad numbers, excluded scope, does not have the proper equipment, or colluded with other bidders.

If all of the bids are too high and significantly above the project's budget, developers may reject all bids or wait for the acceptance period to expire. They can then rebid the project in anticipation of bids from new contractors. However, if it is a public project, the original bids are publicly available, which can influence the next round of bids. Developers may also work with the design team to modify the construction drawings and specifications prior to rebidding to help reduce costs from the original design.

3.8. PERMITS

After site plan approval, some site work activities will require obtaining a permit from the local jurisdiction. Typically, a jurisdiction will require permit applications for environmental impacts, demolition, buildings, land disturbance, stormwater, and any connection to or work within a public facility such as roadways or utilities. Permit applications will vary by jurisdiction and permit type, but often require project information, specific plan sheets from the site and/or building plan, and an engineer's estimate of the permitted work. Permits may also take several weeks or a month to obtain depending on the permit type and issuing agency. Each permit will have an associated fee to cover permit administration and inspections during construction. An example of a permit application is shown in Fig. 3.7.

The local jurisdiction may also require the developer to post a bond prior to the release of the permit. In this case, the bond covers the cost of public improvements such as stormwater management and erosion and sediment control should the developer default on all or part of the contracted work. The developer will acquire a bond for the required amount from a bonding company for a nominal fee. For example, if a 1 million dollar bond is required for road improvements, the developer may pay $10,000 to obtain the bond. The bonding fee will vary based on the size of the project and bonding capacity (maximum amount of surety credit) of the developer. A developer with a high bonding capacity working on a large project will have a lower percentage bond fee than a developer with a low bonding capacity working on a smaller project. The jurisdiction typically requires a bond estimate from the civil engineer, and bond payments are held until the work is substantially complete or until the end of the certification period. An example of a certification period is landscaping required to be maintained and healthy for the first year after project completion. Should the work remain incomplete for a predetermined period, the locality can use all or a portion of the bond payment to complete the work. Bonds are separate from permit fees which are nonrefundable, regardless of the status of completion. If a bond is not required for a permit, the local jurisdiction will require the developer to post the bond before construction commences.

The contractor is generally responsible for obtaining construction permits based on approved plans and all developer paperwork being on-file with the permit-issuing agency. The civil engineer may assist the developer in preparing the permit applications including associated paperwork and following up with the review authority to facilitate

Permit Application for Land Disturbing Activities

<table>
<tr><td rowspan="5">Applicant or Contractor</td><td colspan="3">Individual or Company Name</td></tr>
<tr><td>Name of Representative</td><td colspan="2">Title of Representative</td></tr>
<tr><td>Phone No.</td><td>Fax No.</td><td>Email Address</td></tr>
<tr><td>Address</td><td colspan="2">City, State, Zip Code</td></tr>
<tr><td colspan="3">Contractor's License No.</td></tr>
<tr><td rowspan="2">Property Owner</td><td>Name</td><td colspan="2">Phone No.</td></tr>
<tr><td>Address</td><td colspan="2">City, State, Zip Code</td></tr>
</table>

Street name & address(es) of proposed work: ______________________

____

Purpose of work or activity:

Construction		**Demolition**
☐ New Residential	☐ Clearing/Grading	☐ Commercial
☐ New Commercial	☐ Excavation/Fill	☐ Multi-family
☐ Building Addition	☐ Landscaping	☐ Single-family
☐ Driveway/Parking Lot	☐ Other: _______	☐ Tree Removal
		☐ Other: __________

Full Description of Work

____

____

____

Total area of land disturbance: ______ square feet, ____ acres

Is any part of this property located within a floodplain? ☐ Yes ☐ No

Responsible Land Disturber (RLD): *Certification No:* *Phone No.:*

____

I certify I have the authority to complete this application, the information in this application is correct, and the work will comply with all state laws and ordinances, rules, regulations, policies, and special conditions of the County.

Signature of Applicant: *Date:*

____

Print Name: *Phone No.:*

____

Note: Work covered under this permit must be started within 6 months of permit issuance, and this permit will be suspended if work is suspended for more than 6 months after beginning work.

FIGURE 3.7 Example permit application.

and expedite the review. One common exception are environmental permits obtained from federal or state agencies where the civil engineer or another specialized consultant will apply for the permit on the developer's behalf as an authorized agent. A developer can obtain a permit or post any required bond at any time between site plan approval and construction commencement, whereas contractors will not obtain any permits until the developer awards the contract. If a developer obtains the permits, it can reduce the delay between awarding the contract and start of construction. Any permits obtained or bonds posted by the contractor are reimbursed by the developer through the construction contract. It is important to specify in the bid documents which permits and bonds contractors are responsible for, so that they can include these numbers in their bid.

3.9. TRAFFIC MANAGEMENT

A traffic management plan will be required for any work within a right-of-way, including roadway improvements, shoulder closures, and/or road closures to facilitate utility installations. Oftentimes, the civil engineer will include the standard temporary traffic control details from the Federal Highway Administration's (FHWA) Manual on Uniform Control Devices (MUTCD) in the permit drawings. If extensive road closures and/or detours are required, a traffic engineer may prepare a project-specific traffic management plan.

The contractor may prepare traffic management plans beyond what is included in the permit drawings for additional work in the right-of-way or through coordination with the onsite inspector. These additional traffic plans may be prepared by the contractor or may require the civil engineer's or traffic engineer's involvement. The contractor will also have to coordinate the timing of any lane or road closures with the inspector and local jurisdiction, including scheduling around special events such as peak tourist seasons or time of day restrictions. Depending on the local jurisdiction, additional traffic management plans and scheduling of closures may be required before the start of construction.

In addition to a vehicular traffic management plan, pedestrian access must also be considered during construction, particularly maintaining accessible routes. The civil engineer may include pedestrian detours as part of the traffic management plan in the permit drawings, but the contractor may also have to modify the pedestrian access routes based on field conditions and sequencing of construction activities.

3.10. PRECONSTRUCTION MEETING

A preconstruction meeting is typically required by the local jurisdiction at the project site. The contractor is required to set up this meeting a predetermined number of days prior to construction commencement. Representatives from the authorities having jurisdiction over stormwater, urban forestry, erosion and sediment control, utilities, and traffic will typically attend this meeting to discuss the project with the contractor. All permits and bonds should be in place prior to this meeting. The locality's representatives will likely discuss the construction schedule and any additional permissions or notifications required for specific construction activities, such as road closures, noise ordinances, or disruption to utility service.

In addition to the required meeting with the local jurisdiction, it is recommended to have a meeting between the developer, design team, and construction team between the notice to proceed and construction commencement. This will allow all parties to review procedures during construction, such as shop drawing review, RFI procedures, and progress meetings. This initial meeting will initiate further progress meetings during construction as discussed in Chapter 5. The developer may wish to invite any third-party inspectors and representatives from the local jurisdiction to this preconstruction meeting in addition to any required follow-up onsite meetings.

CASE STUDY: LIQUIDATED DAMAGES

ISSUE:

Enforceability of liquidated damages of $700/day.

SUMMARY:

The case arises from a modest public works project for construction of a traffic signal and related intersection improvements, at a stipulated price of $683,300. Based on excerpts in the appellate decisions, the construction contract appears to include EJCDC® C-520 or similar. The village inserted "$700/day" in the liquidated damages clause governing unexcused contractor delays in completion.

Partly as the result of subcontractor problems, and perhaps because of site difficulties, the contractor was 397 days late in completing the work. The contractor made weak attempts at seeking additional time and compensation, but never complied with the contract's formal notice provisions. At the $700/day rate, the total damages for late completion were liquidated at $277,900.

The contractor filed a lawsuit against the village seeking to enforce the contractor's claims for additional time and compensation. The village countered with a demand for enforcement of the liquidated damages clause. The trial court ruled in favor of the village on both issues, and the contractor appealed. The Court of Appeals held that the contractor's claims should be rejected, primarily on procedural grounds.

The contractor had not followed the contract's procedural requirements for claims, and did not properly appeal a differing site condition ruling.

Somewhat surprisingly, however, the Court of Appeals ruled in favor of the contractor with respect to liquidated damages. The court held that liquidated damages of more than a third of the total contract price was an unenforceable penalty. The court noted that there was no evidence presented regarding the legitimacy of the $700/day amount. There were no supporting calculations, and no relevant background facts such as a record of accidents at the intersection. The court mentioned that the intersection had never previously had a traffic light, so the lengthy delay merely sustained the status quo.

The liquidated damages issue was appealed to the Ohio Supreme Court.

DECISION:

The Supreme Court of Ohio reversed the Court of Appeals, and issued a strong decision in favor of the enforceability of liquidated damages provisions on public works projects. Among the court's conclusions favoring liquidated damages (LDs):

- By agreeing to LDs, the parties avoid controversy over the amount of damages.
- An LD clause is an advance settlement of damages from a future breach in meeting the completion deadlines.
- The law should not look with disfavor on LD provisions (citing a 1948 S. Supreme Court decision).
- LD clauses promote prompt performance of contracts.
- If the LD amount is a "genuine covenanted pre-estimate of damages" that is not a punishment for default, it is enforceable.

- It is uniquely difficult in public works construction to calculate damages to the public interest, making it advisable to agree to damages in advance.
- Delays in public projects result in inconvenience, increased costs, and loss of use to the public.
- LD clauses that establish a rate (daily, weekly) are especially favored, compared with lump-sum LD clauses.
- The Court of Appeals wrongly focused on the total amount of LDs—the focus should have been on the reasonableness of the per diem amount.
- The LD rate must be reasonable, but does not need to bear an exact relation to actual damages.
- The analysis of an LD clause must be prospective (looking forward from the point when the contract was formed, based on facts known at that time) rather than retrospective (focusing on aggregated damages after the fact).

COMMENT:

The Ohio Supreme Court made some pointed comments in its decision. It labeled the Court of Appeals fixation on the aggregate damages as "myopic." It concluded that looking at the aggregate amount would lead to a "perverse rule of law" that would result in short delays being assessed liquidated damages, but not more egregious lengthy delays. The intermediate court was also chided for not enforcing the plain terms of the contract, but rather undertaking "to be wiser than the parties."

As noted in 2015, the dispute does raise a worthy point in asking what basis the owner had for the $700 rate. The Ohio Supreme Court decision supports the decision by pointing out many published public works cases in which the daily rate was similar. Nonetheless, EJCDC strongly encourages owners (and their engineering consultants) to establish a daily damages rate using reasonable criteria, and to carefully document the reasoning. Such documentation in the case might have carried the day at the Court of Appeals level, despite the large total.

It is important to note that there was no criticism in the case of the EJCDC liquidated damages clause's wording or structure.

As a footnote, the case includes a citation to an 1822 U.S. Supreme Court case on liquidated damages for delay. It is interesting to learn that the contract drafters of the early 19th century were already using liquidated damages provisions.

Case study based on Boone Coleman Construction, Inc., v. Village of Piketon, Ohio. *Supreme Court of Ohio (2016).*

The information provided does not constitute legal advice. Consultation with an attorney is strongly advised.

3.11. CONCLUSION

The bidding process described in this chapter largely follows the process for a publicly owned project awarded to the lowest bidder, but as discussed, the bidding process can vary greatly depending on the developer, project type, and project delivery method. The process outlined in this chapter also assumes permit drawings are approved prior to bidding the project. In an effort to expedite the project, the developer may choose to bid the project with permit drawings that are 60–90% complete, such that the bidding overlaps with design completion. In this situation, any modification to the permit drawings after the bid opening may necessitate a change order from the contractor to adjust the cost and schedule. The civil engineer will also need to track specific changes from the drawings used for bidding and approved permit drawings to coordinate with the developer and contractor.

However the contractor is selected, the bidding documents prepared to bid a project will be referenced throughout construction, and therefore it is important that the civil engineer and the developer are familiar with the outlined requirements. In the case of a public project, it is paramount for the developer and design team to follow the prescribed processes, communicate with bidders, and create a transparent selection process.

CHAPTER 4

CONSTRUCTION COST ESTIMATES

4.1. INTRODUCTION

As discussed in the Land Development Handbook, a civil engineer will typically prepare cost estimates throughout the design development phases, including feasibility cost estimates and preliminary cost estimates. A civil engineer may prepare estimates to include with the project's permit or bond applications, typically based on unit cost data provided by the local jurisdiction and quantities derived from the site plan.

The civil engineer will also prepare construction cost estimates after approval of the permit drawings to replicate anticipated contractor bids. Construction cost estimates use relevant pricing and market data and should be prepared by a civil engineer familiar with the construction process, means, and methods. A cost estimate's level of the detail will vary with the detail of the design; therefore, the construction cost estimate is the most detailed estimate prepared by the civil engineer.

4.2. CONSTRUCTION ESTIMATES

Construction cost estimates, also referred to as detailed estimates, engineer's estimates, or opinions of probable cost (OPC), are generally developed only after the local jurisdiction approves the site plans. However, a developer may ask the civil engineer to complete a quantity take-off of the site plans prior to site plan approval. Therefore, the civil engineer should date all take-offs and estimates as well as reference a dated plan sheet. This ensures that the final construction cost estimate is based on the final approved plans used for construction.

Construction cost estimates can be performed by different professionals but should always be reviewed and approved by experienced professionals. This is to assure thoroughness and credibility of the estimate. Furthermore, a construction manager or contractor should perform a constructability review prior to construction commencement and either develop an estimate or check the estimate prepared by others. It is also important for the estimator to be able to visualize the various construction phases of a project and to consider both equipment and materials.

A civil engineer should reference the latest available cost information when preparing a construction cost estimate. Publicly bid projects often publish recent bid information on materials and labor. Later in this chapter, the Sources of Unit Cost Data section provides a listing of possible sources for this information. However, when using a published source, it is often by its nature a snapshot in time resulting in static unit prices and subsequently cost. Construction materials are often subject to fluctuation in prices of raw material such as steel, fuel, and liquid asphalt, and therefore adjustments are necessary as commodity prices rise and fall for these materials even over a short period of time. For example, steel prices may rise 20% within a six-month period. Therefore, the engineer should include a budgeted range for unit prices subject to raw material cost fluctuations.

The developer and civil engineer should avoid using significantly aged data when possible. This can include infrastructure record drawings in the local jurisdiction's files, previous surveys, previous geotechnical reports, and old cost data for previous projects of similar scope and scale. The civil engineer should recommend the developer acquire more up-to-date information. Some of the information can be verified through new surveys and site visits. Other information, such as types and quantity of soil and rock, fiber optic cables and gas lines, can be more difficult or costly to verify without excavating the site. For any cost estimate, the contingency percentage should be adjusted based on the level of detail of the available information and the status of the design documents.

In general, the construction estimate should include the quantities and unit costs of all items needed to construct the project. Similar to preliminary estimates, the construction cost estimate can be used by the developer for loan applications, cash flows, and development of market prices. Additionally, the developer may use this estimate to determine the amount of funding needed to complete the project.

After the plans are complete, measurement of construction materials can be more accurately quantified. The design team can use the approved site plans to verify preliminary estimates prior to bidding or release of construction documents. This cost estimate should still include a contingency to accommodate unforeseen site conditions that may be uncovered during construction. A typical contingency for a construction cost estimate is 10% but can vary based on the information available including reliability of cost data. Contractors also prepare an estimate in order to bid on the project. If the civil engineer's and the contractors' estimates differ significantly, the reason should be understood. A poor economy may prompt contractors to bid low because of the need for work, but low bids may indicate a poor understanding of the work involved. If contractors' bids come in high, the developer may want additional contractors to bid on the project or confirm the design team has not made errors in their estimate. As an example, sometimes sharp price fluctuations in steel or crude oil may significantly affect the unit price of items such as structural steel or asphalt.

The civil engineer should include a standard disclaimer on all estimates. This disclaimer should state the cost estimate is based on the information available and engineering judgment with similar projects. The civil engineer also has no control of the cost of material and labor or market conditions as noted previously. The civil engineer should also not guarantee the estimate's accuracy relative to bids or ultimate project cost.

Often the developer will ask contractors for bids on selected portions of the project, including infrastructure, grading, and roadway construction. Therefore, the civil engineer's construction cost estimate should be formatted so that portions of the estimate can be directly compared with partial bids. When comparing several bids, wide disparities in the bids usually indicate a lack of clarity in the plans, specifications, or the instructions to the bidder. One or two marked deviations from all of the bids, however, usually indicate some oversight or lack of understanding by the bidder.

In general, the engineer's estimate should be accurate within 5–10%. The accuracy depends on the complexity of the plans and specifications and how current the bid data used to develop the unit costs is. Additionally, errors in the design plans can greatly affect the accuracy of the cost estimate. Back charges and extras by contractors due to incomplete and/or inaccurate plans will increase costs significantly.

Since construction cost estimates utilize quantity take-offs from construction drawings, the accuracy of the cost estimate is dependent on the accuracy of the quantities and estimated unit prices. In addition, the civil engineer must recognize that multiplication of numbers, as performed when applying unit cost to quantities, can result in magnification of errors. Equation 4.1 shows the propagation of errors in the product of two independent variables, where A and B are the independent variables and ε is the uncertainty or error in the variables. For example, the estimate for excavation is 10,000 CY (A) ± 20% (ε_A) and the unit cost is \$10.80/CY (B) ± 10% (ε_B). Using Equation 4.1, the accuracy of the cost (C) is:

$$C \pm \varepsilon_c = (A \times B)\sqrt{(A \pm \varepsilon_B)^2 + (B \pm \varepsilon_A)^2} \quad (4.1)$$

$$(10{,}000 \pm 20\%) \times (\$10.80 \pm 10\%) \quad (4.2)$$
$$= (10{,}000 \pm 2{,}000) \times (\$10.80 \pm \$1.08)$$
$$= (10{,}000 \times \$10.80)$$
$$\pm \sqrt{(10{,}000(\$1.08))^2 + (\$10.80(2{,}000))^2}$$
$$= \$108{,}000 \pm \$24{,}150$$
$$= \$108{,}000 \pm 22.4\%$$

Note that although the estimated range (i.e., error) in quantity is 20% and the estimated range in the cost is 10%, the overall accrued error (ε_C) is 22.4%

When two independent variables are added together, the accrued error is represented by Equation 4.3:

$$C \pm \varepsilon_c = (A + B) \pm \sqrt{\varepsilon_A^2 + \varepsilon_B^2} \quad (4.3)$$

Table 4.1, generated from Equation 4.1, provides the expected error given the errors of the two independent variables. For example, the error of 22.4% calculated with Equation 4.2 can be obtained by entering 20% for variable A and reading across to the 10% column of variable B. Table 4.2 shows how errors are accrued when numbers are multiplied and added together. Construction cost estimates need to be carefully reviewed to minimize errors as they can compound throughout the estimate.

It is imperative that the civil engineer includes every item in the construction drawings in the construction cost estimate for equitable comparison with contractor bids and estimates. It is especially important for the construction cost estimate to provide an allowance for contingent items such as rock excavation, pipe bedding, excavation and removal of unsuitable material, undercutting, and other similar items. These items need to be identified in order to allow the developer to budget for them accordingly as either an allowance or part of their contingency. Allowances should be separated from the quantifiable lump-sum items.

It is recommended that engineers utilize a standard checklist and a statement cost estimate sheet or develop ones if they do not exist. A typical checklist is shown in Fig. 4.1. A standard cost estimate sheet is provided in Appendix C.

Table 4.3 shows a procedural list for a civil engineer to follow when preparing a cost estimate. It is intended to provide a framework for developing a cost estimate, but not all items will be needed for every estimate.

TABLE 4.1 Percent Error Expected in the Product of Two Independent Variables

	Percent Error in Variable *B*						
% Error in Variable *A*	**0**	**5**	**10**	**15**	**20**	**25**	**30**
0	5	5.0	10.0	15.0	20.0	25.0	30.0
5	5	7.1	11.2	15.8	20.6	25.5	30.4
10	10	11.2	14.1	18.0	22.4	27.0	31.6
15	15	15.8	18.0	21.2	25.0	29.2	33.6
20	20	20.6	22.4	25.0	28.3	32.1	36.1
25	25	25.5	27.0	29.2	32.1	35.4	39.1
30	30	30.4	31.6	33.6	36.1	39.1	42.5

TABLE 4.2 Compounding of Errors Showing Quantities Times Unit Costs

Item	Quantity	Unit Cost	Total Cost	Accuracy ($)	Accuracy (%)
Excavation	10,000 CY ± 20%	$10.80 ± 10%	$108,000	$24,150	±22%
Concrete sidewalk	250 LF ± 15%	$40.00 ± 20%	$10,000	$2,500	±25%
36-inch-diameter concrete pipe	320 LF ± 10%	$90.00 ± 15%	$28,800	$5,192	±18%
			$146,800	$31,842	±21.7%

4.3. VARIABLES AFFECTING COST

Many variables affect the cost of a project including:

- size of the job
- location of the site
- season
- current state of economy
- financial factors
- cost of property
- type of the zoning and use
- any utility or road relocation
- any unusual structures(s) (i.e., retaining walls, bridges)
- unforeseen site or subsurface conditions (i.e., rock excavation)
- inflation factor.

4.3.1. Size and Complexity of the Project

The size and complexity of the project can have a significant effect on the cost of physical improvements, particularly when considered on a cost per unit basis. Smaller sites can require a disproportionate cost for infrastructure improvement, particularly with respect to streets and underground utilities. Economy of scale, or quantity discounts on construction materials, may not apply for smaller sites.

4.3.2. Site-Specific Conditions

Development costs can vary greatly due to site-specific conditions. As an example, if public water, gravity sanitary sewer and/or storm drain systems are not readily available, extending these services or providing onsite treatment can greatly increase construction costs for a project. Additionally, roadway improvements including traffic signals, turn lanes, or roadway widening can involve substantial construction costs and can have a serious impact on the per acre or individual lot development cost.

If earthwork balance cannot be achieved, fill must be borrowed from an offsite source or excavation and embankment material must be exported. Earthwork import or export is typically at least double the cost of onsite earthwork, and therefore, errors in estimating earthwork can significantly impact the total site-work cost. Use of proprietary software packages which allow for the compilation of quantities directly from the civil engineering design files increases the accuracy of earthwork quantity takeoff through comparison of digital terrain models (DTMs).

The construction cost estimate should also consider any buffers required to protect adjacent neighborhoods using extensive landscaping, sound berms, or walls. These are typically required on sites that border less intense uses such as single-family residential and are required of both rezoned and by-right sites. If possible, these costs should be estimated and included within early feasibility cost estimates.

4.4. SOURCES OF UNIT COST DATA

The accuracy of the cost estimate depends heavily on the accuracy of the unit cost data in addition to correct quantities and contingencies. Published information, such as R.S. Means, is a good source

SUBDIVISION NAME ______________________________ (Section, Lot No.)

Contract/Work Order No. ______________________ Date ______________

____
Signature

***All cost estimates shall be prepared by a senior engineer who will include all the following items in the cost estimate. (If not, briefly describe why!)

1. ______ **Clearing and grubbing**

2. ______ **Excavation:**
   - a. ______ Class A, Rock excavation
   - b. ______ Class B, Regular
   - c. ______ Class C, Wet
   - d. ______ Class D, Haul
   - e. ______ Class E, Borrow
   - f. ______ Class F, Shrinkage

3. ______ **Sheeting and shoring**

4. ______ **Utility lines**
   - a. ______ Sanitary sewer lines
   - b. ______ Storm sewer lines
   - c. ______ Underdrains
   - d. ______ Waterlines
   - e. ______ Electric
   - f. ______ Gas
   - g. ______ Lighting
   - h. ______ Remove and replace pavement

5. ______ **Utility lines should be estimated considering the following:**
   - a. ______ Depth of excavation
   - b. ______ Gravel underdrains
   - c. ______ Size and classification of main
   - d. ______ Linear foot quantities
   - e. ______ French underdrains
   - f. ______ Sheeting and shoring
   - g. ______ Hand excavation
   - h. ______ Rock excavation
   - i. ______ Wet excavation

6. ______ **Concrete work**
   - a. ______ Curb and gutter work
   - b. ______ Sidewalks
   - c. ______ Wingwalls
   - d. ______ Cut-off walls
   - e. ______ Riprap
   - f. ______ Reinforced concrete
   - g. ______ Miscellaneous concrete work
   - h. ______ Retaining walls

7. ______ **Pavement**
   - a. ______ Subbase
   - b. ______ Base
   - c. ______ Paving - Asphalt/concrete
   - d. ______ Paving - Portland cement

8. ______ **Seeding**

9. ______ **Sodding**

10. ______ **Driveways and driveway aprons**

11. ______ **Guardrails, reflectors**

12. ______ **Painting**

13. ______ **Lighting**

14. ______ **Engineering**
   - a. ______ Engineering design
   - b. ______ Surveying stakeout

15. ______ ***Point out in cover letter items not included, such as:**
   - a. ______ Land costs
   - b. ______ Interest costs
   - c. ______ Inspection costs
   - d. ______ Builder's overhead
   - e. ______ Legal fees
   - f. ______ Onsite work
   - g. ______ Planning and preliminary engineering work
   - h. ______ Advertising and public relations
   - i. ______ Permits and bonds.

16. ______ **Riprap**

17. ______ **Gravel - various kinds**

18. ______ **Structural steel work**

19. ______ **Timber construction**

20. ______ **Temporary construction**
   - a. ______ Access roads
   - b. ______ Diversion ditches
   - c. ______ By-passes

21. ______ **Silt control**
   - a. ______ Seeding
   - b. ______ Silt basin

22. ______ **Contingency item (10%)**

FIGURE 4.1 Cost estimate checklist.

TABLE 4.3 Procedure for the Preparation of Cost Estimates

All cost estimates shall be prepared by a civil engineer familiar with the project who shall, prior to beginning any quantity and cost estimate, undertake the following:

1. Secure the plan that is to be used as the basis for the estimate, and *clearly reference* a set of the plans as the source material upon which the estimate is predicated.
2. Research files and discuss with experienced engineers to determine what jobs, if any, have been done in the vicinity of the subject tract.
3. Obtain soil maps or reports of the tract under consideration, if available.
4. Conduct site inspection by physically walking the site, not just driving by or reviewing aerial imagery.
5. Use a standard cost estimate checklist and itemize each quantity or clearly state why the quantity is not defined.
6. Secure unit costs for all improvements from a reliable source (reputable contractor, current bids, contractor's data report, recent cost estimates).
7. Verify that costs secured are installed/finished construction prices.
8. Check all prices with a experienced engineer to obtain concurrence on prices used.
9. Prepare a chart-form cost estimate detailing type of improvement, quantity, units, unit cost, and total cost.
10. Have addition and multiplication cross-checked by a qualified person.
11. Prepare a cover letter to accompany all cost estimates, making the cost estimate an integral part of the letter (i.e., number of all pages—1 of 4, etc.) to prevent separating the cost estimate from the cover letter that qualifies the estimate.
12. The cover letter should enumerate that only those items listed in the cost estimate were considered. Additionally, it should detail those items that were not considered and why.
13. All cost estimates, regardless of how they are prepared, should qualify soil conditions and their possible effect on development costs.
14. All cost estimates must clearly define the source base plan used for the estimate.
15. All cost estimates must show the preparer's initials or name and the date the estimate was prepared, irrespective of closing signature.
16. All cost estimates should be checked by an experienced engineer prior to being sent to the client.
17. All cost estimates should clearly state that costs are based on present-day dollars.
18. All cost estimates should have rounded totals as final figures, which should include a 10 percent contingency factor added in at the end of the estimate.
19. All cost estimates should state the source of unit costs applied to the quantities.
20. Prepare the estimate in the order of construction events.
21. Include a base plan with the cost estimate and letter.

of unit cost data for a preliminary estimate prepared during the schematic design or development of permit drawings. However, a construction cost estimate should reference more accurate sources of unit cost data such as bid tabulations from recent projects in the same region. The civil engineer should adjust unit prices as needed for inflation and geographical variability. Inflation adjustment is discussed in the Engineering Economics section, and geographic factors to adjust cost are available from online databases or R.S. Means.

When providing unit cost data for an estimate, the civil engineer should check first with the developer to determine if a relationship has been established with a reliable local contractor. If one exists, the contractor can provide information on recent bid data on similar projects, which is the best method of estimating unit costs. Alternatively, the civil engineer may have connections with a local contractor, who can provide unit cost data. In some areas, local construction companies form trade associations to evaluate common problems, promote their work, and publish listings of recent successful bids. Many state Departments of Transportation publish recent bid tabulations from which general information on unit costs can be obtained. The civil engineer should review the quantities and adjust for economies of scale when comparing any bid tabulation to the current project.

Local public works agencies also can provide information on recent costs. Adjustments to this cost data may also be required as the construction costs for a public project may be different from privately constructed projects. An example of bid tabulation from a local agency is shown in Fig. 4.2. In addition, many local governments require performance bonds for private projects and publish an associated unit price list to be used for the bond estimate. An example is shown in Fig. 4.3. Performance bond unit prices may be higher than private construction costs to reflect administration and start-up expenses when the local jurisdiction assumes responsibility for project completion. The bond prices may also be updated infrequently and not accurately represent current pricing conditions. Therefore, these bond unit prices are typically not a good reference for preparing a construction cost estimate requested by the developer.

4.5. ENGINEERING ECONOMICS

As the construction period does not always immediately follow design, the developer may need to consider incorporating various economic factors into the construction cost estimate. To aid in this analysis, the civil engineer should adjust the estimate for factors such as inflation or interest rates using engineering economics equations. As shown in Example 1, even a small inflation rate over a short duration can have a significant impact on total project cost due to compounding interest.

Less commonly, the civil engineer is asked to perform financial calculations to assist the developer in value engineering decisions. For these analyses, the civil engineer will utilize a combination of the interest formulas included in the following section. To complete this analysis, a civil engineer can use a spreadsheet program such as Microsoft Excel or reference engineering economics textbooks.

4.5.1. Interest Formulas

While texts on engineering economics use different symbols, the following are the ones most commonly used and are taken from Principles of Engineering Economy (Grant et al. 1990).

Symbols

i = interest rate per interest period
n = number of interest periods
P = present sum of money
F = sum of money at the end of n periods. It is equivalent to P with interest i added
A = end of period payment or receipt in a uniform series continuing for n periods

Formulas. The following information is provided for use with the interest tables when analyzing engineering economics problems.

$$F = P(1+i)^n \tag{4.4}$$

$$P = F\left[\frac{1}{(1+i)^n}\right] \tag{4.5}$$

$$A = F\left[\frac{1}{(1+i)^n - 1}\right] \tag{4.6}$$

			Company A		Company B		Company C		Company D		Company E	
DESCRIPTION	QTY	MEAS.	UNIT PRICE	TOTAL	UNIT PRICE	TOTAL	UNIT PRICE	TOTAL	UNIT PRICE	TOTAL	UNIT PRICE	TOTAL
Clearing & Grubbing	1	LS	2,000.00	2,000.00	2,500.00	2,500.00	3,280.00	3,280.00	2,000.00	2,000.00	1,000.00	1,000.00
Construction Stakeout	1	LS	2,500.00	2,500.00	1,500.00	1,500.00	1,000.00	1,000.00	2,000.00	2,000.00	1,500.00	1,500.00
Maintenance of Traffic	1	LS	2,500.00	2,500.00	5,000.00	5,000.00	3,000.00	3,000.00	4,000.00	4,000.00	5,000.00	5,000.00
Temp. Concrete Barrier for M.O.T.	140	LF	10.00	1,400.00	15.00	2,100.00	12.00	1,680.00	20.00	2,800.00	40.00	5,600.00
Traffic Drums	50	EA	25.00	1,200.00	85.00	4,250.00	55.00	2,750.00	30.00	1,500.00	50.00	2,500.00
Temp. Pavement Tape Marking	1,020	LF	1.40	1,428.00	1.60	1,632.00	0.60	612.00	1.50	1,530.00	2.00	2,040.00
Class 1 & 2 Excavation	638	CY	15.00	9,570.00	18.00	11,484.00	20.00	12,760.00	22.00	14,036.00	39.00	24,882.00
Select Borrow	100	CY	15.00	1,500.00	5.00	500.00	5.00	500.00	22.00	2,200.00	10.00	1,000.00
Silt Fence for Sediment Control	600	LF	1.50	900.00	1.50	900.00	0.01	6.00	1.00	600.00	2.00	1,200.00
Test Pit Excavation Cont.	40	CY	50.00	2,000.00	10.00	400.00	35.00	1,400.00	50.00	2,000.00	5.00	200.00
Stabilized Construction Entrance	50	SY	25.00	1,250.00	11.00	550.00	11.00	550.00	15.00	750.00	10.00	500.00
Strawbales / Sandbags (Sed. Control)	50	EA	5.00	250.00	5.00	250.00	5.00	250.00	1.00	50.00	3.50	175.00
Temp. Curb Inlet Protection	3	EA	200.00	600.00	200.00	600.00	125.00	375.00	150.00	450.00	75.00	225.00
Select Borrow for Trench Bckfill	25	CY	40.00	1,000.00	5.00	125.00	22.00	550.00	24.00	600.00	10.00	250.00
15" RCP Class V	87	LF	35.00	3,045.00	83.00	7,221.00	60.00	5,220.00	35.00	3,045.00	60.00	5,220.00
18" RCP Class V	14	LF	50.00	700.00	87.00	1,218.00	55.00	770.00	60.00	840.00	60.00	840.00
Bituminous Concrete Base	162	TN	45.00	7,290.00	45.00	7,290.00	50.00	8,100.00	70.00	11,340.00	50.00	8,100.00
Bituminous Concrete Surface	126	TN	47.00	5,922.00	50.00	6,300.00	55.00	6,930.00	70.00	8,820.00	60.00	7,560.00
Aggregate Subbase for M.O.T.	50	TN	50.00	2,500.00	15.00	750.00	15.00	750.00	10.00	500.00	10.00	500.00
Calcium Chloride for M.O.T.	2	TN	250.00	500.00	300.00	600.00	350.00	700.00	50.00	100.00	200.00	400.00
Bitumin. Conc. (Cold Mix) M.O.T.	25	TN	50.00	1,250.00	45.00	1,125.00	50.00	1,250.00	50.00	1,250.00	30.00	750.00
Pavement Profiling (Milling)	552	SY/I	3.00	1,656.00	4.00	2,208.00	5.00	2,760.00	5.00	2,760.00	4.50	2,484.00
Poly. Paving Fabric for Bit. Conc.	120	SY	10.00	1,200.00	1.75	210.00	2.00	240.00	3.00	360.00	2.00	240.00
5" Conc. Sidewalk & S/W Ramps	185	SY	27.00	4,995.00	24.00	4,440.00	27.00	4,995.00	20.00	3,700.00	22.50	4,162.50
		TOTALS		57156		63153		60428		67231		76328.5

FIGURE 4.2 Bid tabulation example.

ARLINGTON COUNTY
DEPARTMENT OF PUBLIC WORKS
PUBLIC IMPROVEMENT PERFORMANCE BOND ESTIMATE

SUBDIVISION/SITE PLAN NAME ____________________

ADDRESS/LOCATION ____________________

SUBMITTED BY ____________________ DATE ____________

Item	Quan.	Unit	Unit Price	Cost	Item	Quan.	Unit	Unit Price	Cost
Excavation including Clearing & Grubbing		Cu. Yds.	$23.75		Trees on Public R/W		Each	$330.00	
Conc. Curb & Gutter		Lin. Ft.	$15.00		Tree Grates		Each	$700.00	
Conc. Drive Entrance		Sq. Yd.	$35.00		Street Lights Type - Thoroughfare		Each	$6,000.00	
4" Conc. Sidewalk		Sq. Yd.	$25.50		Street Lights Type - Colonial		Each	$4,500.00	
Brick Paver Sidewalkon Aggregate Base		Sq. Yd.	$46.25		Restoration: Sod, etc.		Lump Sum		
Brick Paver Sidewalk on 4" Conc. Base		Sq. Yd.	$60.50		Erosion/Silt Controls		Lin. Ft.	$6.00	
Brick Paver Sidewalkon 6" Conc. Base		Sq. Yd.	$70.00		CB-2 Catch Basin		Each	$2,275.00	
Brick Paver Crosswalkon 6" Conc. Base		Sq. Yd.	$140.00		Catch Basin (misc.)		Each		
Wheel Chair Ramps		Each	$315.00		CB-2 Top Remove & Replace		Each	$1,225.00	
Bike Trail 4" Asphalt with 6" Comp. Subbase		Sq. Yd.	$17.00		MH-1 Storm Sewer Manhole		Each	$2,525.00	
Subbase - C.B.R. 30		Cu. Yd.	$29.00		MH-2 Storm Sewer Manhole		Each	$3,300.00	
B-3 Bituminous Concrete Base		Ton	$57.00		Storm Sewer Manhole (misc)		Each		
S-5 Bituminous Concrete Top		Ton	$60.00		MH-1 Top Remove and Replace		Each	$1,000.00	
Remove Conc. Curb & Gutter		Lin. Ft.	$7.00		MH-2 Top Remove & Replace		Each	$1,250.00	
Remove Conc. Sidewalk		Sq. Yd.	$5.00		15" R.P.C. Cl. III Pipe		Lin. Ft.	$44.00	
Remove Conc. Drive Entrance		Sq. Yd.	$6.00		18" R.P.C. Cl. III Pipe		Lin. Ft.	$48.00	
Conc. Retaining Wall		Cu. Ft.	$19.50		24" R.P.C. Cl. III Pipe		Lin. Ft.	$63.00	
Stone Retaining Wall		Cu. Ft.	$12.50		____ R.P.C. Cl. III Pipe		Lin. Ft.		
Storm Water Detention System		Lump Sum							

FIGURE 4.3 Bid tabulation example.

Item	Quan.	Unit	Unit Price	Cost	Item	Quan.	Unit	Unit Price	Cost
4" Water Line D.I.P. Class 52		Lin Ft.	$45.00		8" San Sewer Pipe Less than 10' Deep		Lin Ft.	$55.00	
6" Water Line D.I.P. Class 52		Lin. Ft.	$51.00		8" San Sewer Pipe Greater than 10' Deep		Lin Ft.	$160.00	
8" Water Line D.I.P. Class 52		Lin. Ft.	$55.00		15" San. Sewer Pipe Less than 10' Deep		Lin. Ft.	$80.00	
12" Water Line D.I.P. Class 52		Lin. Ft.	$64.00		15" San. Sewer Greater than 10' Deep		Lin. Ft.	$210.00	
16" Water Line D.I.P. Class 52		Lin. Ft.	$78.00		4' I.D. San. Sewer M.H.		Vert. Ft.	$300.00	
6x6 Tapping Sleeve and Valve		Each	$2,200.00		5' I.D. San. Sewer M.H.		Each	$350.00	
8x8 Tapping Sleeve and Valve		Each	$3,150.00		San M.H. Adj. to Grade		Each	$350.00	
12x12 Tapping Sleeve and Valve		Each	$4,150.00		Tunneling 24" Carrier Pipe		Lin. Ft.	$750.00	
16x12 Tapping Sleeve and Valve		Each	$4,068.00		Traffic Signalization*		Lump Sum		
24x12 Tapping Sleeve and Valve		Each	$8,050.00						
2' Blow Off Valve		Each	$1,500.00						
Fire Hydrant		Each	$1,325.00						
Fire Hydrant (relocate)		Each	$1,000.00						

*Contact the Department of Public Works Traffic Signal Engineer on 358-3575 for cost if required by this subdivision/site plan.

ADDITIONAL ITEMS

____

____

SUB-TOTAL __________

20% ADMINISTRATIVE, ENGINEERING & MISC. FEE __________

TOTAL QUOTE __________

REVIEWED AMOUNT APPROVED BY ARLINGTON COUNTY (FOR COUNTY USE ONLY __________

BY____________________ DATE____________________

FIGURE 4.3 *(Continued)*

$$A = P\left[\frac{i(1+i)^n}{(1+i)^n - 1}\right] \qquad (4.7)$$

$$F = A\left[\frac{(1+i)^n - 1}{i}\right] \qquad (4.8)$$

$$P = A\left[\frac{(1+i)^n - 1}{i(1+i)^n}\right] \qquad (4.9)$$

Engineering economics textbooks will provide tables of factors, which vary by interest rate and number of periods. The factors are usually written as (factor, i, n). For example, given a future value (F), the present worth factor for a 7% interest rate (i = 7) over 10 years (n = 10) is written as (PWF, i, n) or (PWF, 7%, 10) and is equal to 0.5083. This factor is then multiplied by the future value to determine the present value. Alternatively, the present value can be solved using Equation 4.5.

EXAMPLE 1

A civil engineer prepares site plans for a middle school and a construction cost estimate for the developer. The construction cost estimate for site work is \$1,500,000. However, the school board may not have funding for this project for another three years and requested a revised estimate to reflect costs in three years.

Current inflation rates are 1%. Using Equation 4.5, the future value of the project as:

$$F = P(1+i)^n = \$1{,}500{,}000(1+0.01)^3$$

$$F = \$1{,}545{,}452 \qquad (4.10)$$

Delaying the construction by three years results in a 3.03% increase in the construction cost estimate.

CASE STUDY: BIDDING MANIPULATION

ISSUE:

Consequences of manipulation of unit price bids.

SUMMARY:

On a water main extension project, the successful low bidder bid $0.01 per cubic yard (cy) on excavation of rock. The bidder believed that 1000 cy, the "indeterminate quantity assumed for comparison of bids," was too high and represented an opportunity to gain an advantage in the bid competition (the bidder also inflated certain other unit prices). In fact, the contractor encountered and excavated 2524 cy of rock. In multiple claims the contractor contended that unit costs were $190 – $220/cy and asked for an equitable adjustment.

The contractor's claim relied primarily on the contract's differing site condition clause (which the court noted was required by Massachusetts statute). The project owner and the local superior court rejected the claim, resulting in the appeal to the court of appeals.

DECISION:

The court of appeals affirmed the lower court's ruling against the contractor. The court noted that conditions did not differ from those stated in the contract documents: the bidding documents had expressly warned that the 1000 cy number was strictly for bid comparison, and that the actual quantity of rock was "indeterminate." The use of a unit price approach for rock removal was a reasonable way of addressing the uncertainty. Putting aside the quantity factor, the court stated that the nature of the rock itself, and the means and cost to remove it, did not differ in any way from what was anticipated in the contract documents. The court held that the contractor could have protected itself by reasonably approximating the unit cost of removal, and that it "defies logic to invoke 'equity' as a basis for adjustment" under the circumstances.

COMMENT:

The decision does not discuss whether the owner could have or should have rejected the bid based on the plainly manipulative one cent bid for excavation. EJCDC® C-200 allows for rejection of "unbalanced" bids, and this bid appears to fall in that category. It is especially significant that in a competitively bid unit price contract the award rests on the stated comparison quantities—if one bidder ignores those quantities, the playing field is no longer level and the bid is arguably nonresponsive.

Case study based on Celco Construction Corp. v. Town of Avon. *Massachusetts Appeals Court (2015).*

The information provided does not constitute legal advice. Consultation with an attorney is strongly advised.

4.6. CONCLUSION

Construction cost estimates are the most detailed estimates prepared by the civil engineer and rely on the accuracy of unit cost data, quantities, and contingencies. The civil engineer should maintain a record of past project's cost information including bids to constantly refine unit cost data. Additionally, frequent review and coordination with local contractors will aid in updating this information. The construction cost estimate should explicitly account for each quantity of materials being tabulated, emphasizing any assumptions made, particularly with allowances for unknown quantities such as rock excavation or undercut.

Chapter 5

Construction Administration

5.1. INTRODUCTION

During construction, the civil engineer may be responsible for reviewing shop drawings, answering requests for information (RFI), and working through construction issues with the contractor, the developer, and other stakeholders. The developer may also rely on the civil engineer for assistance in reviewing the contractor's change order proposals and pay applications and making frequent site visits to observe the construction progress. Depending on the scale and type of project, the civil engineer may also attend regularly scheduled progress meetings where the development team will review construction progress versus the construction schedule. This chapter defines each of these construction administration responsibilities for a civil engineer and typical procedures. These responsibilities differ from the construction manager, who is often a third party hired by the developer to oversee the contractor and construction, direct the contractor as the developer's designated representative, and coordinate directly with the developer and the development team.

5.2. FRANCHISE UTILITY COORDINATION

As franchise utilities such as gas, electric, and communications are operated by private companies rather than the local jurisdiction, they follow a different design and review cycle than public utilities such as water and sanitary sewer. Design of public utilities is required as part of the permit drawings and the local jurisdiction's approval process. The civil engineer may choose to show demolition or relocation of existing franchise utilities as well as design of new service lines on the permit drawing set, but the local jurisdiction does not review or approve this design. The civil engineer will often coordinate the demolition, relocation, and/or design of franchise utilities with the appropriate company during the development of permit drawings, but typically these utility companies prefer to delay their work and design until construction begins. While the construction contract may transfer responsibility of franchise utility coordination to the contractor, the civil engineer plays an integral role in the ongoing franchise utility coordination. The developer may ask the civil engineer to review easement plats and design documents from the franchise utility company to confirm the proposed alignments do not conflict with any proposed site components including public utility lines, storm drains, landscaping, or walls.

5.3. SHOP DRAWINGS

Shop drawings (also known as submittals or cutsheets) are details, diagrams, calculations, and/or product data prepared for design components not fully detailed in the construction drawings to demonstrate conformance with the construction drawings and specifications. Shop drawings can include standard specification sheets from a manufacturer as shown with the manhole shop drawing in Fig. 5.1 or a site-specific drawing such as braces for steel beams and columns. The contractor may also

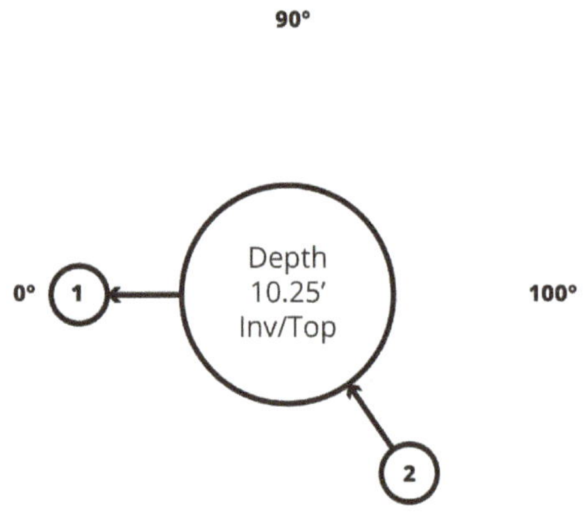

Plan View

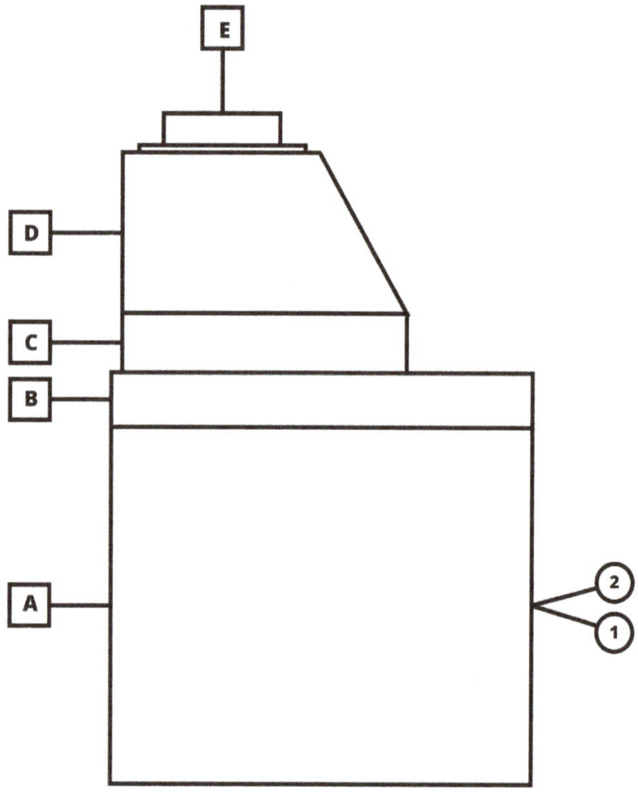

Structure Details
Joint Type: Offset
Joint Sealer: Mortar
Step Angle: 118°
Sump: 0.44′

Elevations
Top: 320.89′
Top of Base Slab: 310.20′
Btm of Base Slab: 309.54′

Elevation View

Opening Details

Tag	Pipe	Invert Elevation	Opening	Angle	*Vertical Location
1	42″ RCP-C	310.64″	58″	0°	0″
1	42″ RCP-C	310.84″	58″	205°	2″

* Reference point is top of base slab - distance is to bottom of opening (or center if marked "CTR")

Stacking Details

Tag	Build Height	Description	Weight (lbs)
A	72″	5-4 Reg Base [72]	10000
B	10.75″	Flatop Reducer [72 x 48]	3100
C	12″	1-0 Riser (Hoop) [48]	1000
D	32″	2-8 Cone [48 x 24]	2700
E	8″	MH-1 "STORM" F&C	300

Weights are approximate.

Customer: Omni Corp.
SO#
Eng: BM
OCP
Project: Delta City
Detroit — Michigan
Structure Type: Manhole
Date: 11/5/2015 8:51:07 AM
Structure: 1

FIGURE 5.1A Example shop drawings.

ABC Stone Co.

Washington County, CA

Size Aggregate: Class 1, California Department of Transportation Standard Specifications

% Retained on #4 sieve	Dry Wt lbs/cu ft	Optimum Moisture %
30	147.0	6.3
35	148.7	5.9
40	150.5	5.6
45	152.2	5.2
50	154.0	4.9
55	155.9	4.6
60	157.8	4.2
65	159.7	3.9
70	161.7	3.5

Absorption of +4s (5)	0.5
Maximum Density −4s (lbs/cu ft)	137.7
Optimum Moisture −4s (%)	8.3
Specific Gravity +4s	2.80

Figure 5.1B Example shop drawing.

include material samples for the civil engineer's review. The shop drawing can also be submittal of aggregate distribution as shown in Fig. 5.1A and 5.1B.

The contractor typically compiles shop drawings and associated information following the notice to proceed and submits to the design team for review. While construction activities such as site demolition may begin prior to approval of shop drawings, the design team will need to approve the shop drawings prior to the contractor ordering those materials. The contractor, or the developer's construction manager, may separate the shop drawings per discipline so that the civil engineer only receives site-related items to review. Alternatively, the contractor may send the entire set of shop drawings to one member of the design team, likely the prime consultant, for distribution. Regardless, the civil engineer should only review site-related items and not provide any comments or stamp any shop drawings received unrelated to site-work items.

The civil engineer reviews the submitted shop drawings for general conformance with the contract documents including the site plan and specifications. Oftentimes, site plans will specify a particular manufacturer and model number, and on public projects, any reference to one manufacturer is typically followed by "or approved alternative." For example, the plans may specify a "Megalug or approved equal" for the mechanical joint restraint required on a waterline. Contractors may submit a Megalug detail and specification sheet as the shop drawing, or they may submit an equivalent restraint device from another manufacturer. When the contractor submits a different manufacturer and product, the civil engineer references a local

SHOP DRAWING REVIEW

☐ NO EXCEPTIONS ☐ EXCEPTIONS NOTED

☐ REVISE AND RESUBMIT ☐ REJECTED

REVIEWED ONLY FOR GENERAL CONFORMANCE WITH THE PROJECT REQUIREMENTS INDICATED IN CONTRACT DOCUMENTS AND FOR CONSISTENCY WITH THE PROJECT DESIGN CONCEPT. THIS REVIEW DOES NOT RELIEVE THE CONTRACTOR FROM RESPONSIBILITY, FOR ERRORS OR OMISSIONS IN DESIGNS FOR WHICH THE CONTRACTOR IS RESPONSIBLE, FOR COMPLIANCE WITH ALL REQUIREMENTS OF THE CONTRACT DOCUMENTS AND FOR THE SAFE AND SUCCESSFUL CONSTRUCTION OF THE WORK. THIS REVIEW DOES NOT CONSIDER THE MEANS, METHODS, TECHNIQUES, SEQUENCES AND OPERATIONS OF CONSTRUCTION, OR SAFETY, PRECAUTIONS OR PROGRAMS INCIDENTAL THERETO, WHICH ARE THE SOLE RESPONSIBILITY OF THE CONTRACTOR.

BY: ______________________ DATE: ______________

ABC ENGINEERING, LLC

WASHINGTON, D.C.

Figure 5.2 Shop drawing stamp.

jurisdiction's list of approved products if available or compares the specifications for the product designated in the site plans versus the product selected by the contractor.

Typically, a civil engineer will have a shop drawing review stamp such as the one shown in Fig. 5.2. The stamp typically includes four options: *no exceptions, exceptions noted, revise and resubmit,* and *rejected,* as well as a disclaimer detailing the limits of the review. *No exceptions* (or *no exceptions taken*) indicates the shop drawing is in conformance with the site plans and specifications. *Revise and resubmit* designates there are modifications required to the shop drawing, and the civil engineer requests an opportunity to review those changes. A civil engineer may use this when multiple revisions to the shop drawing are required or the revision required is critical to meeting the site plans and specifications. Similar to *revise and resubmit, exceptions noted* indicates modifications required to the shop drawing, but the civil engineer does not need to review the revisions. In this case, the contractor incorporates the civil engineer's notes into the product selection. Examples of when a civil engineer may use *exceptions noted* include updating a dimension on a shop drawing or clarifying that a manhole lid labeled "STORM" may be used for both storm or sanitary sewer applications, provided the label is altered accordingly. A civil engineer may reject a shop drawing when it does not meet the contract documents and cannot be modified to achieve conformance. A product may also be rejected if a local jurisdiction does not allow the product to be used.

The contractor develops a submittal schedule and shares this schedule with the design team. The civil engineer or the developer's construction manager should enforce this schedule as to not receive all shop drawings at one time or in an illogical sequence, either of which could lead to confusion and possibly delays. The contractor may also make a delay claim as a result of late shop drawing review or processing. As shop drawing review and approval is a time-sensitive matter due to long lead times for some material manufacturing and delivery, a civil engineer should review any unapproved shop drawings with the contractor as soon as possible. In addition to detailed notes on the shop drawing, the civil engineer should explain to the contractor why revisions or a different product are necessary to help avoid multiple iterations of shop drawing review. To expedite the shop drawing review process, contractors should not submit shop drawings when they select an item on the local jurisdiction's approved product list or specified on the plans. Civil engineers should return shop drawings like these to contractors without reviewing in order to maintain their construction administration scope and subsequently manage their fee.

Additionally, a civil engineer is generally not liable for any construction errors resulting from shop drawings that do not meet the plans and specifications. Even if the civil engineer reviewed the shop drawing and indicated no exceptions taken, as stated in the disclaimer on the shop drawing review stamp, the civil engineer did not approve the error, and it is the contractor's responsibility to build the project in accordance with the plans and specifications.

5.4. REQUEST FOR INFORMATION

Contractors submit request for informations (RFIs) to the design team when they need further clarification of the construction documents or specifications. The RFIs may also request a product substitution or solicit the design team's input on a proposed field change to the design. Contractors submit RFIs during the bidding phase of the project as well as during construction. They will prepare a written RFI with the request and indicates when they need a response in order to avoid impacting the construction schedule as shown in Fig. 5.3. They are also responsible for recommending a cost-effective and compliant solution. Following this procedure minimizes or eliminates a potential change order, and therefore, the civil engineer should return any RFI that does not include a recommended solution.

The civil engineer or appropriate design team member should review the RFI and submit a written response back to the contractor. Some RFI responses will be written clarifications while others will necessitate a sketch or plan revision to the plans to correct a design error. Any changes to

REQUEST FOR INFORMATION

Project:	Main Street Retail	RFI No.:	12
	Reston, VA	From:	ABC Contractor
To:	Dewberry	RFI Date:	43467
	8401 Arlington Blvd	Project Number:	123-45
	Fairfax, VA 22031		

Subject:	Storm Manhole Detail
Response Requested by:	1/8/2019
Drawing / Specification Reference:	CG - 101 & CG - 501 - Structure A1

RFI Description

The storm and grading plan specifies structure A1 is to be a brick and mortar manhole constructed in the field. CG-501 only provides a detail for a pre-cast manhole. Is either structure type acceptable?

Attachments:	n/a
Submitted by:	John Smith

RFI Response:

Either the brick and mortar manhole (per VDOT Std. MH-1 detail) or precast manhole (per VDOT Std. MH-2) are acceptable for structure A1.

Attachments:	n/a
Response by:	Suzie Jones, P.E.
Date:	1/3/2019

FIGURE 5.3 Example of a RFI with a response from the civil engineer.

the plan during the bidding phases are included in the addenda, whereas any changes during construction should be incorporated into the contractor's as-built drawings and/or the civil engineer's record drawings. The civil engineer should carefully consider RFI responses during construction and any subsequent plan revisions as a contractor may seek a change order to cover additional or unexpected costs from the revisions. Ultimately, it is the civil engineer's responsibility to provide a solution that is in the best interest of public safety, the developer, and the final product. Therefore, ideal solutions avoid a change order while upholding these principles.

If the contractor submits an excessive number of RFIs, the civil engineer needs to raise the concern to the contractor and developer. When the RFIs submitted contain information contained in the construction documents, the civil engineer should return the RFI to the contractor and alert both the contractor and developer that the design team is spending time on inappropriate RFIs, which is creating additional costs to the developer. For a design-build project, the design is not as fully developed and therefore subcontractors and the design-build contractor may be looking to the civil engineer to answer more questions. In this case, the civil engineer should work with the design-build contractor to more effectively manage the number of inquiries from the subcontractors and seek additional compensation as needed.

5.5. CHANGE ORDER

A contractor will submit a potential change order (PCO) to the developer when the project conditions or details are modified, and once approved, the PCO becomes a change order (CO), which updates the contract amount. An example of modified project conditions is encountering additional rock in the existing subsurface, therefore exceeding the rock excavation allowance in the original contract. An example of modified project details is replacing the pavers specified in the contract drawings with stamped concrete. In this case, the developer will expect the change order to be a reduction in the overall contract price, also known as a credit to the developer. A change order will specify changes to the project cost and/or schedule. For changes to the project scope such as including an additional building, the contractor and developer will sign a supplemental agreement rather than negotiating a change order.

Typical causes of change orders include unforeseen site conditions, developer-requested revisions during construction, ambiguous contract documents, or errors and omissions in the construction documents. In all cases, the contractor did not anticipate these conditions during the bidding phase and therefore did not include this work in the original schedule or fee. Typical unforeseen site conditions include unsuitable soils or construction debris, rock, non-located existing utility lines, inspector requirements beyond the approved site plan, dewatering, and any other construction item the developer explicitly excluded from the bid. However, in some cases, developer exclusions from the bid are not eligible for a change order including if the contract or specifications establish an unclassified soil site. With unclassified sites, the contractor is responsible for accounting for onsite soil conditions including excavation in the bid. Developer-requested changes include value engineering decisions like replacing sod with seed and functional or aesthetic site revisions such as changing the surface of a multi-use trail from gravel to asphalt. Ambiguous contract documents, errors, and omissions include discrepancies between detail drawings and jurisdictional standards, mistakes in design, and omitting required features such as parking signs on the site plan. A plan revision issued by the design team that affects the project cost or discovery of altered site conditions will prompt the contractor to prepare a PCO request for the developer's review and approval. A contractor cannot use a change order to cover any bid omissions of items specified on the contract documents. For example, if the site plans included two park benches and a plaza area at the corner of the property but the contractor failed to include these site features in the bid, the requested change order cannot include the cost of these benches and additional hardscape. The contractor is responsible for including all items in the original bid so the developer can ensure all bids received are based on the same project scope.

When preparing the PCO, a contractor will consider new work added to the project, any previously completed work affected by the revision, and the agreed-upon markup. The new materials and labor will often be more expensive than the original contract due to economy of scale. Typically, the contractor will order the original materials in bulk, but with a change order, the supplier will charge a higher unit price for the smaller amount of material. If the revision requires the removal and replacement of completed work such as paving or landscaping, the contractor will include a price for this rework. While the developer should receive a credit for any previously ordered materials not needed for the revised plan, the contractor may reduce this credited amount based on any restocking fees for the returned materials. For these reasons, when preparing a plan revision, the design team should discuss the construction status with the contractor and collaborate to mitigate the amount of installed improvements affected with the revised plan. Additionally, some plan revisions during construction may necessitate a review by the local jurisdiction prior to implementing in the field, so the design team should consider any impacts to the overall schedule as well.

In most cases, developers approve PCOs; however, they may defer to the design team to review and provide comments on the cost. The civil engineer should request a copy of the original construction contract from the developer to aid in the review of the change order. The civil engineer should immediately return any PCO where the contractor provides a reason for the PCO. Otherwise, by reviewing and recommending the developer approve the PCO, the civil engineer accepts the contractor's reason and therefore potentially accepts fault for the PCO. The PCO will include materials, quantities, equipment, and labor, which differs from the generalized unit costs typically found in an engineer's estimate. To review a PCO, civil engineers should review the quantities in the PCO compared to quantities shown on the revised plan. They can also compare any PCO unit prices to the original bid, particularly if the bid included unit prices. Civil engineers should also validate any items for rework against site conditions or recent pay applications. Finally, they should review any items such as unsuitable soils covered under an allowance in the original contract to confirm they are not double-counted with the PCO. They provide comments to the developer, who in turn negotiates the final PCO amount with the contractor.

Developers also reserve the right to evaluate the cause for all change orders. They should include a contingency with the project budget to cover cost overruns and change orders. However, they may also look to the design team or contractor to cover the cost of the change order if it is determined either party is at fault. Litigation may occur when the developer and design team or contractor do not agree who is responsible for paying the change order.

Change orders can be reasonably expected as an aspect of any project. The design team should work with the developer prior to the bidding process to estimate the range of changes and costs that may arise throughout the project. Constructability reviews in the later phases of the design process by a third party can identify potential construction issues and therefore minimize change orders. The design team can address these issues by making changes to the design or possibly identifying the need for the contractor to modify the construction sequence or phasing to avoid a particular issue. Unexpected subsurface conditions, unforeseen utility and drainage work on adjacent properties, and design team- or developer-directed changes are among the most common reasons for change orders. Best practices to avoid surprises and minimize the cost of change orders include effective communication and coordination between the design team and the contractor, reasonable allowances included within the developer's project construction budget, and contract language outlining the expectations and responsibilities of all parties. Regular input and review between the design team and the developer during the design and throughout the construction process can prove to be beneficial to the overall success of the project.

5.6. PAY APPLICATION

A pay application is the contractor's version of an invoice. The contractor typically submits a monthly pay application to the developer for review and payment. The pay application will show the contractor's estimate of percentage complete for each activity such as paving, landscaping, and utility installation. The pay application will provide the scheduled value agreed-upon in the original contract as well as retainage, which is an amount the developer withholds from payment to the contractor until project completion and closeout. Figure 5.4 shows an example of a pay application.

The developer will often forward the pay application to the design team to review against current site conditions. In this case, the design team should request a copy of the original construction contract and any approved change orders, so they can review the pay application against the agreed-upon terms. Upon receiving a pay application, the civil engineer should make a site visit to confirm the percentage complete for each line item. If there is a discrepancy between what the contractor is claiming and what is observed onsite the civil engineer should notify the developer but is not responsible for quantifying the actual percent complete. The contractor revises the pay application based on the civil engineer's comments or provides additional documentation to justify the original pay application such as providing receipts for material delivery onsite.

5.7. PROGRESS MEETING

Typically for public or large-scale projects, the design team, the developer's project manager (or construction manager), and the contractor will meet for regularly scheduled progress meetings. The construction manager typically leads the meeting with input from the developer's project manager on agenda items. If the developer did not hire a construction manager, the prime consultant will often lead these meetings. The purpose of the meeting is to review the current status of construction and any outstanding issues. The contractor will report on completed work and forecast work to be completed prior to the next progress meeting (see Fig. 5.5). The contractor will also report on any outstanding items or issues such as RFIs, shop drawings, or potential change orders. The frequency of the meetings may vary from project to project but will occur at an interval agreed upon by the development team at the beginning of construction.

5.8. CONSTRUCTION SCHEDULE

A contractor prepares a schedule to sequence activities, which are discrete tasks required to complete the project, as well as establish intermediate deadlines. The schedule considers time for material delivery and construction as well as weather days typical for the geographical region. While there are many scheduling techniques a contractor may use, the ultimate objective is to find the critical path. The critical path is the sequence of activities that will take the longest time considering material acquisition and construction duration, and therefore, dictate the overall project schedule. The schedule typically includes some float time, which allows an activity to be delayed without affecting the overall project schedule. For example, a waterline installation will take five days and is scheduled to start on February 1. The waterline installation must be completed before the pavement installation on February 13, so the start of the waterline installation may be delayed until as late as February 8 without impacting the overall schedule. In this example, there are seven calendar days of float in the waterline installation. The critical path does not have any float.

The construction schedule (see Fig. 5.6) is a working document the contractor uses to gauge progress toward project completion and make adjustments in sequencing as needed to obtain the agreed-upon project completion date. A copy of the construction schedule may be provided to the developer and design team for reference and is a useful resource to reference during the progress meetings. The civil engineer may prepare a less detailed construction schedule for the developer's reference prior to site plan approval and awarding the bid. However, once the bid is awarded, the construction schedule becomes the contractor's responsibility to prepare and maintain.

As discussed in more detail in Chapter 7, the critical path for a land development project often includes activities associated with early grading and drainage improvements to prepare the building pad and then initial building construction. Therefore, the contractor will typically schedule any activities necessary to prepare the building pad first, followed by building erection activities while site work continues simultaneously. The contractor will also

ABC Contractor

Project Name: Main Street Retail

Pay Application Number: 6

Pay Application Date: 1/2/2019

Period to: 12/31/2018

Item No.	Description of Work	Scheduled Value	Work Completed		Materials Presently Stored	Total Completed & Stored to Date	% Complete	Balance to Finish	Retainage (10%)
			Previous Applications	This Period					
01	Storm Sewer	$ 15,000	$ 15,000	$ —	$ —	$15,000	100%	$ —	$ 1,500
02	Sanitary Sewer	$ 10,000	$ 10,000	$ —	$ —	$10,000	100%	$ —	$ 1,000
03	Waterline	$ 12,000	$ 12,000	$ —	$ —	$12,000	100%	$ —	$ 1,200
04	Asphalt	$ 30,000	$ —	$ 10,000	$ —	$10,000	33%	$ 20,000	$ 1,000
05	Concrete	$ 15,000	$ —	$ 5,000	$ —	$5,000	33%	$ 10,000	$ 500
06	Landscaping	$ 5,000	$ —	$ —	$2,000	$2,000	40%	$ 3,000	$ 200
07	Signage	$ 3,000	$ —	$ —	$ —	$ —	0%	$ 3,000	$ —
	GRAND TOTAL	$ 90,000	$ 37,000	$ 15,000	$ 2,000	$ 54,000		$ 36,000	$ 5,400

FIGURE 5.4 Example of a pay application.

Main Street Recreation Center—Construction Progress Meeting #3

August 5, 2019, 10:00 AM

123 Main Street
Washington County

Meeting Lead: John Smith, P.E.

Purpose: Construction Coordination Meeting

Attendees: Sarah Davis, Thomas Wright, David James, Kimberly Bryant

Agenda Items

1. PAYMENT APPLICATION
2. SCHEDULE UPDATE
    - Work completed over the past 30 days
    - Anticipated work over the next 30 days
    - Waterline extension – update
3. SUBMITTALS
    - Upcoming submittals
    - Outstanding submittals
4. CHANGE ORDERS
5. OPEN DISCUSSION
    - Road closure – status
    - Coordination with franchise utilities – update
    - Erosion and sediment control inspections – update
    - Neighbor complaints – update

FIGURE 5.5 Progress meeting agenda.

consider seasonal tasks when preparing the construction schedule. For example, any activities such as earthwork and grading that will impact soils above the frost line need to be completed between spring and fall in colder to moderate climates. Similarly, depending on the lift thickness, asphalt cannot be installed at temperatures less than 40–50 degrees, and therefore, the local asphalt plant may close during winter months.

Table 5.1 shows typical durations for site work activities. These are rule of thumb estimates that vary depending on location and site conditions.

5.9. SITE OBSERVATIONS

During construction, civil engineers typically will make site visits to review any site construction issues with the contractor and observe the general progress of construction. The frequency of these site visits will vary with each project and may be regularly scheduled or as requested. As discussed in Chapter 1, they should wear appropriate personal protective equipment including steel-toed boots, a safety vest, and a hardhat as well as observe best safety practices. While onsite, civil engineers typically observe the work, not for the purpose of guaranteeing the contractor's work, but to generally evaluate whether the work appears to be proceeding in a manner that has been specified in the contract documents. Civil engineers are not typically responsible for and should not provide direction or instruction to the contractor regarding any means and methods of construction. Civil engineers' responsibility for site safety may depend on the terms of the contract; however, they should report any site safety concerns observed to the contractor, construction manager, and/or developer so

Commonwealth Center CVS

William A. Hazel, Inc.
4305 Hazel Park Court
Chantilly VA 20151

ID	Task Name	Start	Finish	Duration
0	CVS at Commonwealth Center Landbay B	Mon 3/12/18	Fri 8/10/18	107 days
1	**OWNER REQUIREMENTS**	**Mon 3/12/18**	**Tue 5/29/18**	**55 days**
2	Notice to Proceed	Mon 3/12/18	Mon 3/12/18	0 days
3	Pre-Construction	Mon 3/12/18	Mon 3/12/18	0 days
4	Permits	Mon 3/12/18	Mon 3/12/18	0 days
5	LCSA Approvals	Mon 3/12/18	Mon 3/12/18	0 days
6	Approved Paving Design-Ph 1	Wed 4/18/18	Wed 4/18/18	0 days
7	Approved Paving Design-Ph 2	Tue 5/29/18	Tue 5/29/18	0 days
8	Authorization to Order Materials	Mon 3/12/18	Mon 3/12/18	0 days
9	**EARTHWORK**	**Tue 3/27/18**	**Fri 5/11/18**	**34 days**
10	Earthwork	Tue 3/27/18	Fri 5/11/18	34 days
11	Trucking	Mon 5/7/18	Fri 5/11/18	5 days
12	CVS Pad On-Grade	Fri 4/6/18	Fri 4/6/18	0 days
13	**ROCK BLASTING**	**Tue 4/3/18**	**Thu 4/5/18**	**3 days**
14	Blasting	Tue 4/3/18	Thu 4/5/18	3 days
15	**UTILITIES**	**Mon 3/26/18**	**Thu 5/17/18**	**38 days**
16	Utilities - Ph1	Mon 3/26/18	Fri 4/27/18	24 days
17	Utilities - Ph2	Fri 4/27/18	Thu 5/17/18	14 days
18	**PHASE 1 - CVS Parking Lot Area**	**Wed 5/2/18**	**Wed 6/13/18**	**30 days**
19	**CURB & GUTTER**	**Wed 5/2/18**	**Tue 5/15/18**	**10 days**
20	Ph 1 Curb Stone	Wed 5/2/18	Thu 5/3/18	2 days
21	Ph 1 Curb and Gutter	Tue 5/8/18	Tue 5/15/18	6 days
22	**SERVICES & CONDUIT**	**Fri 5/18/18**	**Tue 5/29/18**	**7 days**
23	Ph 1 Water Services	Fri 5/18/18	Fri 5/18/18	1 day
24	Ph 1 Conduit - (By Others)	Wed 5/23/18	Tue 5/29/18	4 days
25	**ROADS & PAVING**	**Fri 6/1/18**	**Wed 6/13/18**	**9 days**
26	Ph 1 Base Stone	Fri 6/1/18	Thu 6/7/18	5 days
27	Ph 1 Base Asphalt	Tue 6/12/18	Wed 6/13/18	2 days
28	**PHASE 2 - Turnlane & Lower Lot Area**	**Tue 6/12/18**	**Tue 7/31/18**	**35 days**
29	**CURB & GUTTER**	**Tue 6/12/18**	**Tue 7/3/18**	**16 days**
30	Ph 2 Pre-Grade for UD's	Tue 6/12/18	Tue 6/12/18	1 day
31	Ph 2 Underdrains	Fri 6/15/18	Tue 6/19/18	3 days
32	Ph 2 Curb Stone	Fri 6/22/18	Mon 6/25/18	2 days
33	Ph 2 Curb and Gutter	Thu 6/28/18	Tue 7/3/18	4 days
34	**ROADS & PAVING**	**Mon 7/9/18**	**Tue 7/31/18**	**17 days**
35	Ph 2 Base Stone	Mon 7/9/18	Thu 7/12/18	4 days
36	Ph 2 Base Asphalt	Tue 7/17/18	Tue 7/17/18	1 day
37	Asphalt Trail	Fri 7/20/18	Tue 7/31/18	8 days
38	**FINAL TOPPING**	**Fri 8/3/18**	**Fri 8/10/18**	**6 days**
39	Final Topping	Fri 8/3/18	Fri 8/3/18	1 day
40	Striping and Signs	Wed 8/8/18	Fri 8/10/18	3 days

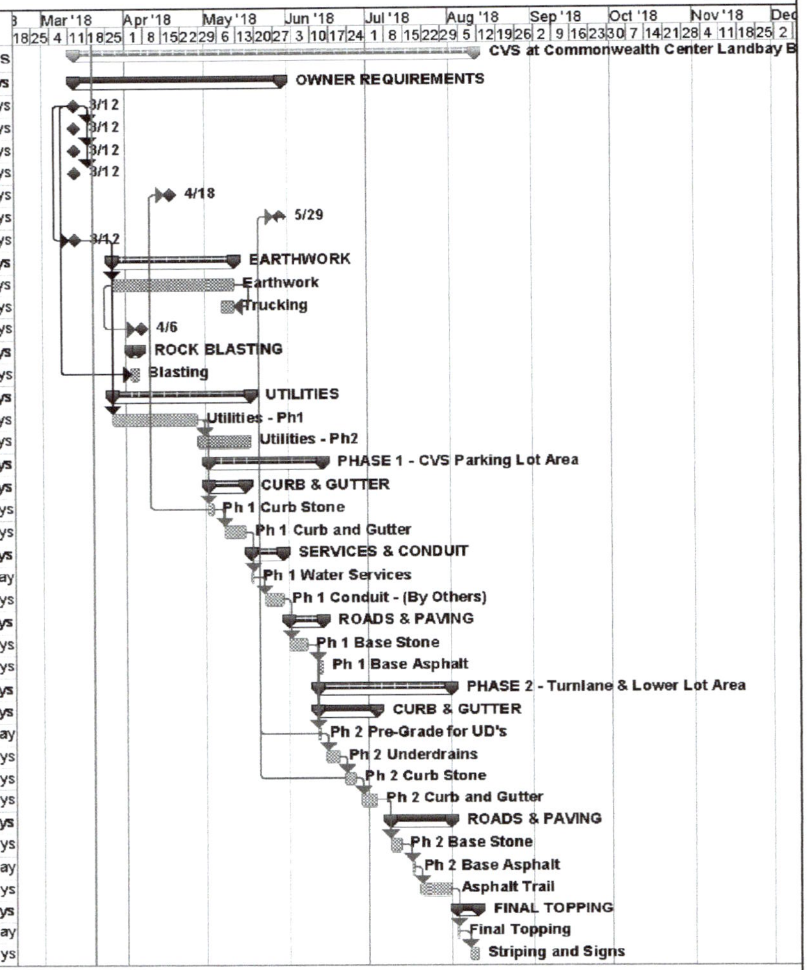

Owner: LF Jennings Inc
WAH Bid No. 18-051
WAH Job No. NYA

THIS SCHEDULE DOES NOT INCLUDE DELAYS DUE TO WEATHER OR OTHERS

Prepared By: NYA
Project Manager: Doug Palatt
Print Date: Tue 3/20/18

FIGURE 5.6 Construction schedule for a commercial project. (Courtesy of William A. Hazel, Inc.)

TABLE 5.1 Typical Durations for Site Work Activities

Activity	Unit	Duration Estimate	Variables
Earthwork & Grading			
Regular Excavation	CY	min. 750 CY per day	• Weather • Number of haul trucks available for use • Location of disposal site
Borrow Excavation onsite	CY	min. 750–1500 CY per day	• Weather • Number of haul trucks available for use
Borrow Excavation offsite	CY	min. 500 CY per day	• Weather • Number of haul trucks available for use • Location of borrow pit
Undercut Excavation	CY	+/– 500 CY per day	• Weather • Number of haul trucks available for use • Location of disposal site • Significantly less production than regular excavation if soil is being evaluated by a Geotechnical Engineer as excavation is occurring
Pavement Items			
Asphaltic Concrete	TN	1300–1400 TN per day BM-25.0 @ 4" thickness Or 4000 LF of a 12' single lane road per day	• Weather • Area to be paved • Entrances or areas requiring hand work will slow production • Number of haul trucks available • Location of asphalt plant • Mechanical issues with paver • Mechanical issues at asphalt plant
Aggregate Base Material	TN	min. 500–1000 TN per day @ 6" depth	• Weather • Number of haul trucks available • Location of material
Geotextile Fabric (Subgrade Stabilization)	SY	2000 SY per day with a small crew	• Number of workers helping roll out/pin down
Flexible Pavement Planning	SY	min. 4000 SY per day (2" thickness)	• Weather • Area to be milled • Areas with limited access will slow production • Size of machine being used • Condition of equipment

(*Continued*)

TABLE 5.1 Typical Durations for Site Work Activities (*Continued*)

ACTIVITY	UNIT	DURATION ESTIMATE	VARIABLES
		DRAINAGE ITEMS	
Reinforced Concrete Pipe	LF	50–75 LF per day	• Weather • Utilities • Crew size • Foreman/crew experience
End Sections	EA	2 EA per day	• Weather • Utilities • Crew size • Foreman/crew experience
Drop Inlets	EA	2 EA per day	• Weather • Utilities • Crew size • Foreman/crew experience
Manhole	EA	2 EA per day	• Weather • Utilities • Crew size • Foreman/crew experience
Box Culvert	LF or CY	Precast: 60 LF per day Cast-in-place slower, multiple days due to pouring the floor, walls, then ceiling	• Weather • Dewatering plan—major variable as most box culvert areas are underwater • Foreman/Crew experience/size
Stormwater Management Excavation	CY	min. 700 CY per day	• Weather • Type of material being removed • Number of haul trucks available • Location of disposal site
		DEMOLITION ITEMS	
Demo of Flexible Pavement	SY	min. 300 SY per day (typically asphalt only)	• Number of haul trucks available • Location of disposal site • Size of excavator being used
Demo of Pavement (combination rigid & flexible)	SY	min. 150 SY per day (typically includes concrete pavement under asphalt)	• Number of haul trucks available • Location of disposal site • Size of excavator being used
		CLEARING ITEMS	
Heavy Clearing (mature trees)	AC	1 AC per day	• Weather • Terrain • Chip & grind

(*Continued*)

TABLE 5.1 Typical Durations for Site Work Activities (*Continued*)

Activity	Unit	Duration Estimate	Variables
Light Clearing (second growth)	AC	2 AC per day	• Weather • Terrain • Chip & grind
Blasting Items			
Diabase Rock	CY	100 CY per day	• Weather • Depth
Silt Stone	CY	200 CY per day	• Weather • Depth
Hoe Ram	CY	20 CY per day	• Areas where blasting is not allowed
Incidental Roadway Items			
Curb (with curb machine)	LF	min. 900–1300 LF per day	• Weather • Crew size
Curb (hand-formed radius)	LF	150 LF per day	• Weather • Crew size
Curb and Gutter (with curb machine)	LF	min. 800–1300 LF per day	• Weather • Crew size
Curb and Gutter (hand-formed radius)	LF	140 LF per day	• Weather • Crew size
Hydraulic Cement Concrete Sidewalk 4"	SY or CY	Less than 200 SY per day or approx. 20 CY per day	• Weather • Crew size
Utility Items			
Water Main (12" DIP)	LF	min. 50 LF per day	• Weather • Crew size • Crew experience • Size of excavator being used • Materials stockpiled onsite
Water Main (6" DIP)	LF	min. 50 LF per day	• Weather • Crew size • Crew experience • Size of excavator being used • Materials stockpiled onsite
Fire Hydrant	EA	1 EA per day	• Weather • Crew size
Sanitary Sewer Pipe (8" DIP)	LF	min. 50 LF per day	• Weather • Crew size • Crew experience • Size of excavator being used • Depth • Presence of rock • Materials stockpiled onsite

(*Continued*)

TABLE 5.1 Typical Durations for Site Work Activities (*Continued*)

Activity	Unit	Duration Estimate	Variables
Manholes	EA	2 per day	• Weather • Utilities • Crew size • Foreman/crew experience • Depth
		Erosion and Sediment Control Items	
Protective covering	SY	min. 500 SY per day	• Crew size
Soil stabilization mat	SY	min. 500 SY per day	• Crew size
Inlet protection	EA	Stone: min. 4 EA per day Gutter buddies: as many as needed	• Weather • Crew size • Materials stockpiled onsite
Temporary silt fence	LF	min. 750–1000 LF per day using trench machine	• Weather • Crew size • Materials stockpiled onsite • Equipment available
		Traffic Signal Items	
Concrete foundation	CY	16–20 CY per day (about 2 pole foundations)	• Weather • Crew size • Crew experience • Equipment being used clamshell excavator vs. slime
Signal pole & mast arms	EA	min. 2 EA per day	• Weather • Crew size • Crew experience • Traffic limitations—scheduling lane closure
		Landscaping Items	
Regular seeding, lime, fertilizer	LB or TN	No limit with hydroseeder	• Weather • Area being treated
Modular block wall	SF	80 SF per day	• Weather • Crew size • Equipment available
Tree planting	EA	Variable	• Crew size • Equipment available
Shrub planting	EA	Variable	• Crew size • Equipment available

that the appropriate party may evaluate and take action to correct any safety issues. Civil engineers often document the site visit with a field report and/or through the preservation of photographs of the observed conditions. After reviewing the field report and photographs with the civil engineer, the developer can follow-up with the contractor to address any issues. A sample site visit report is shown in Fig. 5.7.

5.10. PUBLIC INSPECTIONS

In addition to the civil engineer's site observations and any third-party inspections the developer procures, the local jurisdiction will also require a public inspector to inspect the building construction and some site components throughout construction in order to issue a certificate of occupancy at the end of construction. Site components requiring inspection will vary for each jurisdiction but may include any public utilities such as water or sanitary sewer, stormwater management systems, erosion and sediment control devices, roadway improvements including turn lanes, retaining walls, and landscaping. The contractor will schedule the inspection when the work is complete, and the inspector will make a site visit to review the completed work. Some inspections such as erosion and sediment control inspections occur regularly and are not scheduled by the contractor. Some inspections fees will be covered with the original permit application and fee, whereas other inspection fees may be paid separately. The inspector will note any deficiencies in a report that will need to be addressed by the contractor. Depending on the deficiencies noted, the civil engineer may be involved in designing solutions to address the inspector's comments.

Dewberry Site Visit Report

Project:	North Main Villas
Date:	1/2/2019
Site Visit Conducted by:	Lillian White, P.E.

Site Conditions and Work in Progress

Lots 1, 2, & 3 are nearing completion, pending final stablization.
Construction Entrance for Lot 4 installed and site clearing is complete.
Waterline extension appears complete and ready for pavement overlay.

Site Observations:	
	Waterline extension appears complete. Per contractor, trench will be backfilled next week and paving will be completed by next Friday.
	Construction entrance installed at Lot 4.
	Silt fence in locations throughout the site separated from stakes. Picture taken along northern property line of Lot 2. Recommend developer follow up with contractor to repair/replace as needed throughout the site prior to jurisdiction inspection.

Figure 5.7 Example of a site visit report.

CASE STUDY: APPROVED ALTERNATE

ISSUE:

Subcontractor's obligation to supply specified liner system after engineer's rejection of proposed "or equal."

SUMMARY:

For Des Moines wastewater treatment plant improvement project, project engineer ABC Engineering specified holding tank liner systems by Linabond. Utility-One, a subcontractor specializing in Ameron liner systems, assured Best Construction that Ameron would qualify as an "or equal" and submitted a sub-bid that assumed Ameron. Best Construction relied on this and was awarded the contract. During formation of the Best Construction-Utility-One subcontract, the two exchanged documents and communications about the content of the subcontract. Utility-One wanted to include its bid (which specified the Ameron liner would be provided), whereas the base subcontract merely incorporated the ABC Engineering specs. Ultimately the subcontract that was executed did not include the bid limiting Utility-One's commitment to Ameron. ABC Engineering rejected a request to accept Ameron as "or equal"—reportedly the first rejection of Ameron in 25 years. Utility-One refused to absorb the costs of providing the more expensive Linabond, and contended that it had never intended to provide anything but Ameron. Best Construction pointed out that the subcontract contained no reference to Ameron and no exceptions to the ABC Engineering specification.

DECISION:

The court examined the subcontract's integration clause, stating that no prior communications or agreements were part of the final subcontract, and held Utility-One liable for $241,000 in costs needed to supply Linabond, plus attorney's fees.

NOTES:

There is no discussion in the case as to whether the bidding documents allowed for a prebid "or equal" determination. See EJCDC C-200, Instructions to Bidders, 11.01, second option. The C-700 "or equal" procedures afford a full opportunity to apply for equal status, but ultimately give the Engineer the final decision as to acceptance of a proposed equal. Bidders (prime or sub) proceed at their own risk if they assume a less expensive item is indeed an equal. EJCDC's integration clause is currently built into the definition of "Contract"—C-700, Article 1—and is supported by Article 9 of the Agreement, C-520.

Case study based on John T. Jones Construction Co. v. Hoot General Construction Company, Inc. *United States Court of Appeals for the Eighth Circuit (2010).*

The information provided does not constitute legal advice. Consultation with an attorney is strongly advised.

5.11. CONCLUSION

Each project will be different, and the civil engineer may not be asked to do each of these construction administration tasks during construction. Depending on the developers' experience and in-house staff capabilities, they may ask the civil engineer to be more or less involved in the construction process than discussed above. This may entail more frequent meetings and site visits as well as review of more communication between the developer and contractor. Regardless, the civil engineer and contractor should strive to maintain frequent communication, and the civil engineer should respond to any construction issues expeditiously in order to help the contractor maintain the schedule.

CHAPTER 6

ADVANCED CONSTRUCTION ROLES

6.1. INTRODUCTION

In addition to the construction administration responsibilities described in Chapter 5, civil engineers and other professionals will often hold advanced construction roles. These positions are usually held by individuals that are certified in specific aspects of project development such as construction management, inspections, scheduling, or financing. The scope and responsibilities of these advanced construction roles will vary by project scale and type. The entire development team should be familiar with the larger program operations that may be in place.

Most public and/or complex projects will employ a construction manager (CM) to coordinate between the multiple project teams and oversee project delivery. The CM is usually independent of the primary architect, civil engineer and contractor. In this role, the CM is required to interact and collaborate with the development team and all stakeholders. With some projects, the construction management responsibilities may be included in the scope of services required of the design team members. Construction management requires the respect and cooperation of the entire development team.

All projects will require inspections of specific building and site components to confirm compliance with applicable codes, standards, and specifications. While many of these inspections are administered by the local jurisdiction, the developer may hire a third-party inspector to evaluate private infrastructure. Additionally, the developer may procure a commissioning agent to assist with building commissioning prior to accepting the building from the contractor. Inspectors are involved throughout construction, regularly reporting to the developer and/or local jurisdiction on the status and quality of construction.

This chapter discusses the scope of work associated with construction management and inspections in the construction phase of a project.

6.1.1 Partnering

Partnering is a means of managing the relationship of all parties to enhance the quality, value, and efficiency of a project. Mutual respect for all team members is built on trust, transparency, and collaboration. The Construction Industry Institute (CII) defines partnering as:

> *...a long-term commitment between two or more organizations for the purpose of achieving specific business objectives by maximizing the effectiveness of each participant's resources. This requires changing traditional relationships to a shared culture without regard to organizational boundaries. The relationship is based upon trust, dedication to common goals, and an understanding of each other's individual expectations and values* (1996).

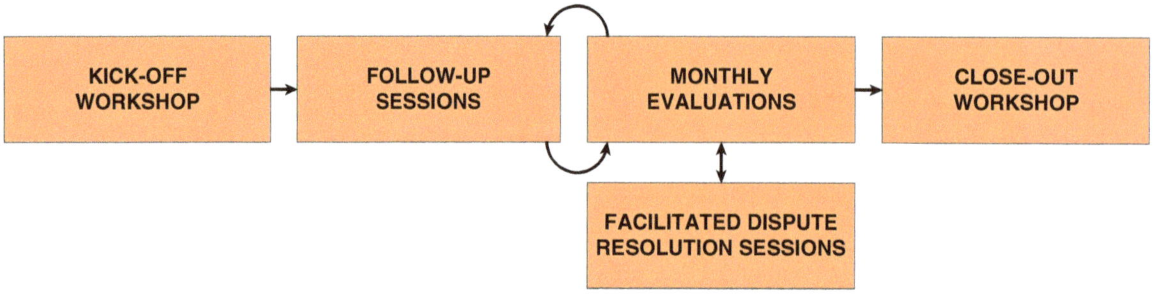

Figure 6.1 Timeline of project partnering.

Partnering promotes communication among key individuals from the development team by making early introductions, allowing each party to express early concerns, and identifying areas that require special attention or nonstandard procedures. All parties are motivated to participate in formal or informal partnering to deliver a successful project. By establishing relationships and issue resolution procedures upfront, everyone agrees on best practices for responding to issues as they arise. Construction disputes between development team members can often be attributed to poor cooperation, lack of trust, and ineffective communication, thereby leading to unnecessary project costs and schedule overruns.

As shown in Fig. 6.1, partnering is a continuous process, often starting with a kick-off workshop where the team develops project goals, a communication plan, and an issue resolution process. Following the workshop, the team then meets regularly (typically on a monthly basis) to review goals, commitments, and emerging issues. If a dispute arises, a facilitated dispute resolution session can mitigate the issue escalating. After project completion, a close-out workshop is held to review partnering performance and document lessons learned.

6.2. CONSTRUCTION MANAGEMENT

Construction management incorporates specialized project management techniques to oversee construction. This management may expand to include oversight during the planning and design phases when a non-traditional delivery method such as design-build or construction manager at risk is used. The construction manager (CM) is responsible for the schedule, scope, budget, and quality in fulfilling the project requirements. This role is typically served by a third-party professional (typically with a background in engineering, architecture, or construction) that is not directly responsible for the design production and is not associated with the contractor. Similar to the licensure and registration requirements for design professionals, a certified construction manager (CCM) is a formal accreditation. Each CCM has professional responsibilities and adheres to a code of ethics.

The role of a CM can be fulfilled by an individual or a team of certified professionals. Careful consideration must be given to staffing the construction management team to balance productivity and cost. The number of staff and their responsibilities on a construction project depends on the complexity of the project, developer requirements, project delivery method, and the number of work areas.

In some cases, the developer or the design team are responsible for some of the construction management tasks. With a small retail project, the civil engineer might be required to provide cost estimation and construction administration services. Alternatively, a highway project likely includes a CM that is responsible for managing cost, safety, schedules, and risk. A developer may also have an in-house professional that is responsible for construction management. The CM may be involved in a project as early as the pre-design phase or may not be engaged until the procurement or bidding phase. The developer should establish each professional's construction management scope of services in the early phases of the project.

Other chapters within this book introduce some of the scope of services associated with construction management, such as cost estimating, safety management, and scheduling. When a CM is involved in the project, there are certain tasks that fall solely within their scope of services. The CM's responsibilities typically include construction quality, budget, schedule, safety, and risk management. Additionally, the CM may be specifically responsible for managing project documents and certifying compliance of sustainability requirements. The Construction Management Association of America (CMAA) *Construction Management Standards of Practice* provides a comprehensive resource for typical responsibilities of a CM. The responsibilities described in the following sections are:

- Quality Management
- Cost Management
- Scheduling
- Green Building
- Documentation
- Building Information Modeling (BIM)
- Issue Resolution

6.2.1 Quality Management

Quality management is fundamental to project delivery success. Quality, as used in this context, refers to a measurable condition of how the project work adheres to the project requirements. The key terms associated with quality management include quality assurance (QA) and quality control (QC). This is often referred to as QA/QC, but each refers to different aspects of quality management. As defined by the Federal Highway Administration (FHWA) in the *Construction Program Management and Inspection Guide*, QA focuses on processes and includes project acceptance, dispute resolution, and inspections to ensure adherence to project requirements. QC focuses on monitoring the project's materials and workmanship as defined in the contract specifications and requirements.

The quality management associated with the construction phase is an extension of the quality management plans produced by the design team during the pre-construction phases. In some project delivery methods, such as construction manager at risk the quality management for the design phase may include CM oversight.

Quality Assurance Plan. The developer, potentially in coordination with the CM, usually drafts the QA plan before commencing construction. The QA plan identifies responsibilities of development team personnel and the qualifications of contractors and subcontractors. It also outlines the protocol for inspections, construction monitoring, design modifications, and project close-out. This protocol typically includes the type, frequency, and required documentation for each activity. Additionally, it establishes a plan for regular project meetings. According to the Transportation Research Board (TRB), a QA plan includes "...elements of quality control (QC), independent assurance, acceptance, dispute resolution, laboratory accreditation, and personnel certification (2013)." The QA plan should consider the planning, design, procurement, construction, and operation phases of the project as well as recognize the relationship between each of these phases.

Quality Control Plan. A QC plan establishes a framework of requirements to guarantee the completed project meets the specified quality requirements. It defines the measurement of various aspects of the quality requirements, compares the work to the contract specifications, and identifies areas of deviation from the desired results, noting suitable modifications where appropriate. The QC personnel are typically part of the CM staff but can be part of the design team or a representative of the developer. It is the QC personnel's responsibility to manage all control activities and tests, prepare acceptable documentation of the QC activities, and develop and maintain an effective QC system.

QC during construction is most often achieved through inspections either by the contractor, a third-party, or inspector appointed by the public review agency. The frequency of inspections, the responsible party, and the desired quality standard are typically established in the QA plan. By monitoring the project, measurable signs of QC issues can be identified before they become critical and impede project success. These inspections and QC monitoring may also be used when evaluating pay application requests or when preparing the final punch lists and closing out the project. A sample QC inspection report is included in Fig. 6.2.

McKenzie CM Services, Inc.

QC Inspection Form

Project:	**Main Street Recreation Center**
Operation:	**Storm Sewer & Storm Drain**
Segment/Location:	**East Parking Lot**
Contractor:	**KW Contractor**

Item	Inspector (Initials)	Date	Notes
Contract Documents Review			
Construction Drawings approved	SOD	4/30/19	approved by Washington Co. 3/15/19
Shop Drawings approved	SOD	4/30/19	reviewed by Dewberry
QA Plan approved	SOD	4/30/19	
QC Plan approved	SOD	4/30/19	
Construction Review			
Verify stakeout	BEM	7/15/19	structure B2 incorrect, adjusted by A1 Surveying 7/12/19
Review trench width and shoring	MAE	7/30/19	trench between B3 & B4 to be completed with Ph 2
Dewatering			
Bedding material placement			
Pipe slope			
Pipe joints			
Structures - inverts, rim elevation			
Backfill placement			
Acceptance			
As-built survey			
Final inspection			

Figure 6.2 QC inspection report.

The inspection phases of a QC plan include:

Preparatory Phase	This phase is carried out before the commencement of any major definable aspect of work. The QC manager organizes preparatory meetings after a thorough review of all applicable plans, specifications, shop drawings, and related submittals.
Initial Inspections	This phase happens at the beginning of a defined aspect of work. It verifies that the control measures developed in the preparatory meeting are implemented and that the level of workmanship appropriate for the work meets the stipulated requirements.
Follow-up Inspections	These inspections, including safety and control testing, are performed daily based on the controls developed during the preparatory meeting and initial inspection. They ensure that products and outputs remain compliant with the contract requirements and that modifications, repairs, and replacement of materials are carried out consistently. Acceptability is verified via reinspection and retesting before additional work takes place.
Final Inspections	These inspections verify the completed work meets the approved plans and specifications. Sometimes these inspections include tests, measurements, and close-out of outstanding items for the project.

6.2.2 Cost Management

The project budget can change as a result of unforeseen issues, design errors, construction errors, schedule delays, requests for information (RFIs) and change orders. The responsibility of managing project cost extends beyond the initial cost estimates and includes initial cost studies, continuous monitoring of project budget, cost projections, and controlling cost changes. In addition to the construction cost estimates prepared by the design team as described in Chapter 4, a third party such as the CM may develop a separate cost estimate and may be involved in reviewing the contractor bids. Throughout the duration of the project, a CM may review pay applications to verify conformance with the appropriate payment request and track project costs with respect to work schedule. Additionally, a CM monitors and documents change orders that increase or decrease the project cost throughout construction to monitor the project budget. Value engineering can be a critical component of cost management that often occurs during the design phase and throughout construction.

Value Engineering. Value engineering is a systematic approach to evaluating the functionality of systems, equipment, facilities, services, and supplies to evaluate construction cost in conjunction with quality, project requirements, and life cycle costs. Value engineering seeks to achieve the highest function when compared to cost. The function can be a measurable performance or a qualitative element of the developer's preference. This process does not necessarily focus on identifying the lowest cost alternatives but instead investigates the value of the design elements. For example, a low impact development (LID) stormwater management system such as a rain garden may have a higher construction cost than an underground manufactured system. However, the LID system may have lower maintenance costs due to no proprietary parts, and additionally, the air quality and aesthetic benefits of the vegetation may yield a higher value to the operator.

During the value engineering process, a professional team will use their individual observation, experience, and knowledge to evaluate a project's design documents. The team will often have knowledge of current material prices, resource availability, constructability requirements, and operational costs that influence the project. Ideally, value engineering occurs in the early project phases because of the inherent challenges and cost associated with making late revisions. Late design changes and rework increase design costs, potentially cause schedule delays, and therefore, decrease the potential value. The design team should continuously monitor the project cost and value during the design phases, but a value engineering evaluation by a third party often provides a new perspective on the design. The value engineering team may suggest alternative design solutions to achieve a better value and will coordinate these suggestions with the design team and the contractor. As a third party, the value engineering team is unlikely to understand all original design requirements and the original context of design decisions. Therefore, many suggestions may not be feasible but provide an opportunity for creativity. Together, the development team can identify feasible alternatives that increase project value.

Contractor Bid Evaluation. Despite the process of formalizing contract and bid requirements to promote consistent bid submissions, there is often a large variation in how the contractor bids are formatted and submitted. The CM may be responsible for reviewing the contractor bids and helping to clarify any potential conflicts prior to issuance of the project contract. Inadvertent exclusions, intentional exclusions, and misinterpretation of the design documents are some of the issues that a CM or other professional may uncover when evaluating contractor bids. Addressing these issues prior to starting the project can protect both the developer and contractor from an erroneous bid submission. As part of the bid evaluation, the developer may ask the CM to prepare a bid tabulation. However, in order to prepare a bid tabulation with individual line items, unit costs, and quantities as shown in Fig. 6.3, the developer must bid the project in a bid tabulation format where the contractor is required to detail individual line items. Otherwise, as with typical lump sum contracts, the contractor may

McKenzie CM Services, Inc.

Bid Tabulation

Project: **Main Street Recreation Center**
Washington County
Parks and Recreation Department

Bid Date: **4/1/2019**

Bid Quantities				Engineer's Estimate		KW Contractor		WC Contracting		Dewey Construction	
Item #	*Description*	*Qty*	*Unit*	*Unit Cost*	*Total Cost*	*Unit Cost*	*Total Cost*	*Unit Cost*	*Total Cost*	*Unit Cost*	*Total Cost*
101	Curb	1,000	LF	$ 15	$ 15,000	$ 15	$ 15,000	$ 25	$ 25,000	$ 18	$ 18,000
201	Drop Inlet	10	EA	$ 3,000	$ 30,000	$ 2,500	$ 25,000	$ 3,000	$ 30,000	$ 3,500	$ 35,000
202	24" RCP	450	LF	$ 60	$ 27,000	$ 80	$ 36,000	$ 65	$ 29,250	$ 70	$ 31,500
301	Fire Hydrant	3	EA	$ 2,000	$ 6,000	$ 3,000	$ 9,000	$ 1,900	$ 5,700	$ 1,500	$ 4,500
302	8" DIP	275	LF	$ 35	$ 9,625	$ 45	$ 12,375	$ 33	$ 9,075	$ 35	$ 9,625
303	8" Gate Valve	5	EA	$ 2,000	$ 10,000	$ 1,700	$ 8,500	$ 1,900	$ 9,500	$ 2,200	$ 11,000
	Total Bid Amount				$ 1,495,000		$ 1,398,000		$ 1,526,000		$ 1,675,000

FIGURE 6.3 Bid tabulation.

break down the lump sum bid into general cost categories such as storm sewer or paving but will not detail the unit costs and quantities. In this case, each contractor may categorize costs differently, thus increasing the difficulty for the CM to compare bids and bid categories equitably.

When available, a detailed schedule of values should also be reviewed during the bid evaluation (see Fig. 6.4). The schedule of values is an itemized breakdown of project elements and can be less detailed than a bid tabulation. The detail and scope of the itemization can vary dramatically between contractors and between projects. Generally, the schedule of values resembles a cost estimate worksheet that includes the item, quantity, and a unit price. The challenge is that the list may be simplified and does not usually include a specific definition of each item. For example, a storm pipe may be listed with a quantity and unit price but the excavation of the trench, installation, inspection, and backfill items might be listed separately. In other cases, each category is listed as lump sum (LS) with no breakdown of quantities. The CM can review the schedule of values and work with the contractor to clarify any ambiguity. During the construction phase of the project, the contractor and developer can negotiate the cost of the change order based on the schedule of values. For example, if additional storm pipe is required to solve a drainage issue, the developer and design team will have a reasonable expectation for the cost of the change order. The schedule of values can also be used to review pay application requests.

KW CONTRACTOR

Schedule of Values

Project: **Main Street Recreation Center**
Washington County
Parks and Recreation Department

Bid Date: **4/1/2019**

Item #	*Description of Work*		*Scheduled Value*
1.01	Bond & Insurance	Labor	$ 15,000
		Material	$ -
2.01	Mobilization	Labor	$ 10,000
		Material	$ -
3.01	Building Demolition	Labor	$ 25,000
		Material	$ -
3.02	Site Demolition	Labor	$ 8,000
		Material	$ -
4.01	Earthwork	Labor	$ 125,000
		Material	$ 25,000
5.01	Erosion & Sediment Control	Labor	$ 7,000
		Material	$ 20,000
	Total Bid Amount		$ 1,398,000

FIGURE 6.4 Schedule of values.

The FHWA *Construction Program Management and Inspection Guide* (2004) provides the following bid review checklist:

1. How many bids were submitted?
2. Is there good competition?
3. How does the bid compare with the engineer's estimate?
4. What is the distribution or range of bids received?
5. Which unit bid prices differ significantly from the estimate?
6. Is there justification for the difference?
7. What is the identity and geographic location of the bidders?
8. What is the potential for savings, if any, if the project is re-advertised?
9. How do the bid prices for the project under review compare to bid prices for similar projects in the same letting?
10. Is there any urgency in letting the project?
11. What are the current market conditions and workload?
12. Are there any unbalanced bids?
13. Would deferral be contrary to public interest?
14. Are there any errors in the engineer's estimate?

6.2.3 Scheduling

Construction involves the complex interaction of people, equipment, manufacturing, and materials. It requires assembling materials in a distinct sequential order based on the physical relationships among components, interactions among trades, unforeseen conditions, specified code regulations, and safety considerations. Therefore, effective construction planning includes appropriate scheduling of the activities necessary to complete the construction project. Scheduling assigns time durations and resources including equipment, materials, and labor to each task. The schedule lists the sequence in which tasks will be completed and is used by contractors to most efficiently allocate resources to the job. A satisfactory schedule minimizes production inefficiency, promotes the prompt acquisition of materials, and defines the critical path to construction completion. Furthermore, an effective project schedule must be accurate for the short term and flexible for the long term, allowing float as needed. Float is the amount of time that a task in a construction schedule can be delayed without causing a delay to subsequent tasks or to the overall completion date. Float in a construction schedule usually allows for adverse weather conditions, manufacturing delays, and production rate losses due to conditions outside of the contractor's control.

Contractors develop a detailed schedule based on the construction start date and substantial completion date established by the developer in the contract. They uses their experience with means and methods as well as typical construction issues to develop an effective schedule, which they are responsible for maintaining throughout the duration of construction. Scheduling methods include the critical path method, bar charts, Gantt charts, line of balancing scheduling, and quantitative scheduling.

One crucial element in scheduling is weather conditions. For example, most horizontal transportation construction is limited to the fall, spring, and summer. The winter is a relatively slower production period because of limitations of earthwork and pavement construction, which contractors must budget for in their schedule. Limitations to the scope of work based on weather conditions should be considered when establishing an initial schedule and monitoring the schedule. For example, a three month delay in the project start may force the scheduled pavement work into the winter months and could actually delay a project by six months if the geographic location experiences harsh winters. Vertical construction can better accommodate adverse weather conditions as the environment can be protected and controlled from impending weather conditions.

While the contractor is ultimately responsible for creating and maintaining the schedule, the CM is often involved in reviewing the schedule, proposing efficient solutions to construction issues, and keeping the developer apprised of the construction status. When revisions to the design documents are required during construction, the CM helps manage prompt responses from both the developer and the design team to mitigate any lost time for the contractor. The CM can help coordinate any submissions to the review agency or modifications to existing permits necessitated by the design change. The CM can also anticipate and address any disruptions to the schedule. One typical disruption to the schedule is weather delays beyond the budgeted weather conditions in the original schedule. The CM and contractor can develop creative solutions to make up lost days and minimize the number of days added to the substantial completion date. If the contractor raises other concerns about the schedule, the CM as the owner's representative can further review the construction status and provide an opinion on the origin of the delay and possible solutions. The CM may be involved in continuous value engineering evaluations as the project is close to acquiring the certificate of occupancy. This value engineering analysis will consider both time and cost such as using off-the-shelf products rather than custom built products to save time. Additionally, the CM may coordinate schedules between multiple parties such as onsite versus offsite work or in the transition between the developer's site work contractor and the homebuilder for a residential subdivision. The developer may also have interim schedule requirements such as tenant move-in for tenant improvements in a retail space that a CM will assist in coordinating.

6.2.4 Green Building

Green building emphasizes sustainability and resiliency within all aspects of a project. In the pre-construction phases of a project, the design team will establish measurement and certification requirements for sustainable practices. Design features like low impact development (LID) stormwater systems, sustainable building materials, efficient appliances, renewable energy systems, and other sustainable elements can promote environmental stewardship of a project. During construction, these green building features should be evaluated and inspected to ensure they meet project requirements. Many green building systems have detailed specifications that will differ from traditional systems and should be given special attention. The construction process itself is also an opportunity to follow green building practices. This can be as simple as the sorting of demolished materials into recyclable materials versus material to be transported to the landfill.

The construction process is a large influence on green building initiatives but is often missed because of the chaotic activity on a construction site. The fluidity of construction activities also presents challenges with properly documenting and managing the daily site activities and disturbance. For example, erosion and sediment control devices and stormwater management systems should be regularly monitored and inspected to ensure compliant operation. Additionally, interim phases of construction may warrant the need for supplemental protection devices that were not anticipated during the design phase. The sequence of construction may also necessitate special protection of LID stormwater systems to avoid heavy construction equipment compacting the system's required permeable soils. Fuel storage for construction activities should also be managed to reduce exposure and risk of environmental contamination.

On a basic level, a job site should be monitored to ensure cleanliness and safety. In early phases of development, if the basics like trash and recycling receptacles are not conveniently available, the job site is prone to littering. Management plans should be established to prevent loose debris within a job site, which can be unintentionally transported offsite. On a larger scale, some projects can benefit from reuse of the demolished site features. Old building materials can be reclaimed and converted to site furnishings, façade treatments, or distributed to other projects or manufacturers that reuse materials. When possible, a project should investigate the opportunity to use local materials to reduce fuel consumption and transport. On a localized level of activity, construction operations within the job site can also contribute to energy consumption that could be otherwise avoided through efficient vehicle and job operations.

Green building rating systems such as Leadership in Energy and Environmental Design (LEED) or Envision typically require documentation throughout the project including construction.

An accredited professional is required to participate in this documentation if project certification is desired. This professional may be part of the design team, construction team, or a third party hired by the developer.

6.2.5 Documentation

The pace and multidisciplinary approach of construction makes effective task tracking challenging and critical. To this end, an effective documentation system captures decisions made by the developer, development team, and/or other project stakeholders. Project documentation is an integral part of construction project management. It ensures that project requirements are fulfilled and establishes a record of tasks including who, when, and what. Documentation provides a permanent record that allows comprehensive reconstruction, review, and analysis of events and actions of the project.

Keeping detailed documentation is one of the most important elements to effectively managing a project. Every construction project is an accumulation of detail, information, and data that must be organized, stored, and distributed clearly, accurately and on time to all project stakeholders. Proper documentation is the basis for payment schedules, legal claims, code compliance, QA and QC, life cycle costings, and maintenance procedures. It aids in the correction of design issues, monitoring of code compliance, efficient tracking of costs, and timely dispute resolution. Having this information organized and readily available is valuable to the developer and eventual operators.

Construction management contracts require written documentation of communication between parties. Clear communication is crucial to the success of a construction project, and written documentation forms the basis for the evaluation and settlement of construction issues. Some examples of vital documents on a construction project include:

- Bidding and contract documents
- Insurance certificates
- Contract specifications
- Drawings
- Bonds
- Shop drawings
- Addenda
- Permits
- Schedule and/or design revisions
- Correspondence
- Meeting minutes
- Pay application requests
- Progress photos and reports
- Punch lists
- Change orders
- Certificates of occupancy
- As-built drawings
- Final acceptance
- Operations and maintenance manuals

6.2.6 Building Information Modeling

Building information modeling (BIM) is the process of design that uses data-integrated-3D object modeling. The models are prepared by the design team, but the digital documents are only one element of BIM. The emphasis is on the process of using the modeled information to improve the design, quality management, construction, and operations of a project. The design models themselves are a revolution of traditional computer-aided design (CAD) drafting techniques, which rely on creating digital shapes to develop the project documents. The modeled elements of the BIM process use intelligent 3D objects with integrated data. For example, a pipe object is represented as a 3D cylinder and will include data like material, size, installation schedule, and cost. BIM promotes the collaboration of design documents and construction operations and supports a multitude of construction program management tasks. During construction, the models can be used to coordinate across disciplines, animate a project schedule, and track resources.

Figure 6.5 depicts a screen capture of Autodesk® Navisworks Manage® software. This software compiles BIM content from various sources such as architectural, MEP, structural, and site plans to create a comprehensive model of the project.

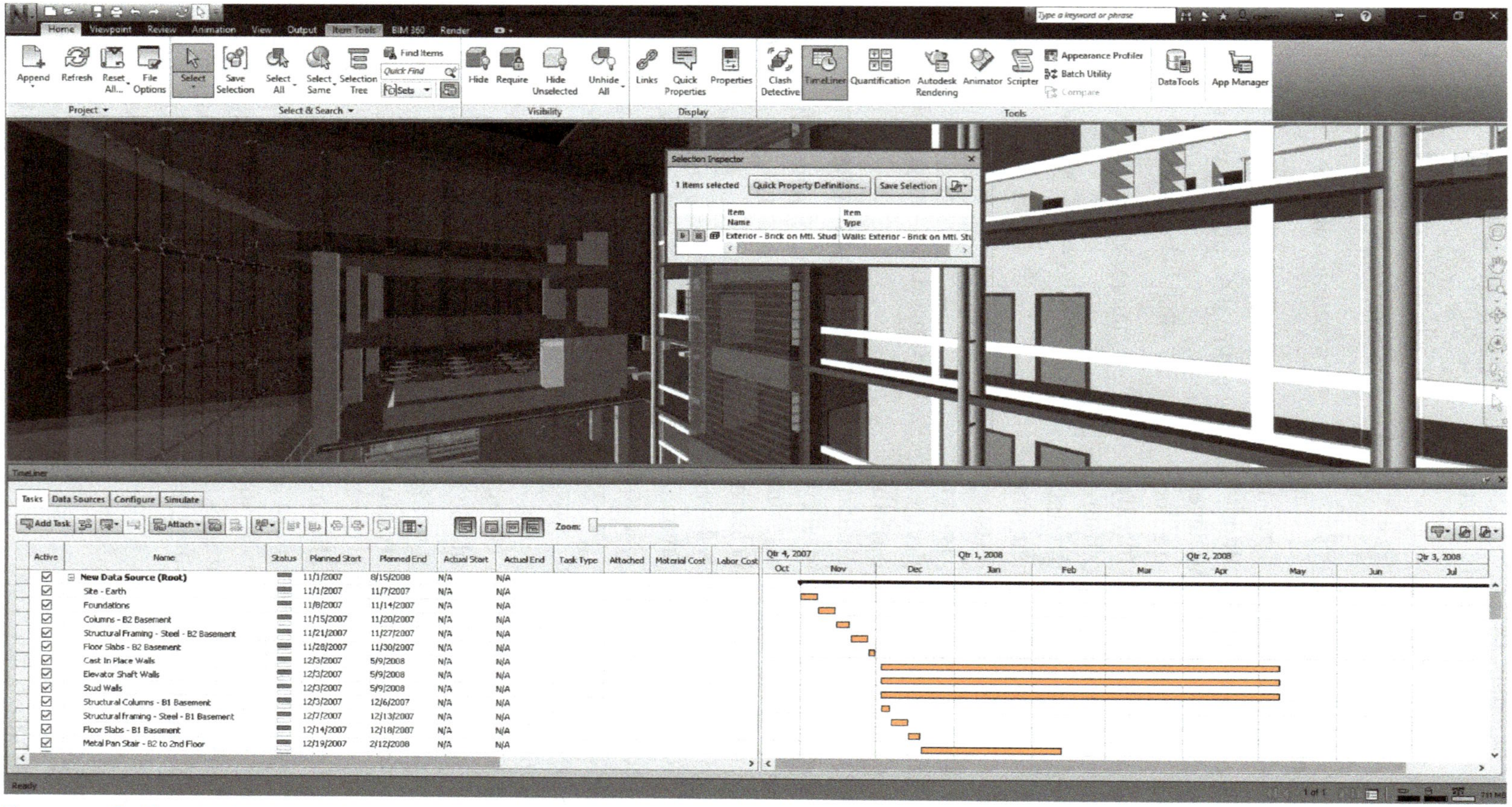

FIGURE 6.5 BIM software, Autodesk® Navisworks Manage®, depicting a 3D BIM view in the upper window and a coordinated schedule below. Model elements are tied to the construction schedule. Autodesk screen shots reprinted courtesy of Autodesk, Inc. Autodesk and Navisworks Manage are registered trademarks or trademarks of Autodesk, Inc., and/or its subsidiaries and/or affiliates in the USA and other countries.

In this view, a schedule presented as a Gantt chart is depicted at the bottom, and the 3D BIM content is visible in the upper window. The schedule can be tied directly to the model components to provide a visual depiction of the construction activities and evaluate potential conflicts.

The models should be managed during construction and updated with material selections and field changes. At the end of construction, the model can be used for asset management by the owner and operators. Many building and infrastructure elements such as electrical systems, plumbing, or site utilities will not be visible after construction is complete, and the model can provide a valuable resource for the facility operators. For example, data such as manufacturer information and installation date can be assigned to an infrastructure element. The facility's management team can monitor the infrastructure through software applications, schedule inspections, and perform preventive maintenance conditions prior to significant maintenance issues or failure. Both the design team and the construction management team are responsible for the continuous updating and management of the model.

6.2.7 Issue Resolution

Issues during construction result from many factors including unsubstantiated or incomplete claims, unclear or vague contract conditions, production delays, failure to administer the contract, poor communication, and resource limitations. Construction management techniques, such as quality management, cost management, and documentation are an effort to mitigate these issues. Situations beyond the control of the project team, such as weather, unforeseen site conditions, and outside influences, may contribute to an undesirable outcome that impacts the project schedule and budget.

The contract's conditions and clauses must be carefully read and understood. Many national organizations, such as AIA and EJCDC, have standardized contracts based on lessons learned in the industry. As discussed in Chapter 3, these standardized contracts can reduce the initial upfront cost of creating a new contract, but with any project, the developer should tailor the contract to the specific project needs. In addition, challenges that arise during the execution of work must be properly documented and dealt with immediately. Project schedules must be realistic and executable with risks identified upfront by all parties involved.

Effectively coordinating with the various personnel involved in the construction process and resolving issues are critical to the success of a project. Resolution of such issues can be achieved with a detailed scope, proper risk assessment, and appropriate resource allocation to avoid significant changes, rework, delays, and cost overruns. Initial planning efforts, responsible management, continuous communication, and risk assessment reduce the frequency of errors.

6.3. INSPECTION AND TESTING

Typically, a construction inspector is required by the local jurisdiction to verify that the contractor performed the work in accordance with the approved plans and specifications. Additionally, an inspector may be hired directly by a developer to evaluate private infrastructure or a portion of the project scope not covered under jurisdictional review. Inspections are performed throughout construction to confirm compliance with the environmental, utility, zoning, building, safety, energy, and state's Department of Transportation (DOT) regulations. Additionally, these inspections need to comply with all federal, state, and local regulations. Specialty inspections may be required to address any threat to the health and safety of adjacent sites, traffic operations, and building users. Materials for structural elements, such as pavement, are tested to certify compliance with construction specifications and regulations including material durability and strength. These tests are usually performed by accredited testing agencies and laboratories. Field- and lab-certified technicians use certified equipment to perform these tests which are reviewed, approved, and sealed by a professional engineer.

On some projects, it is necessary to create an inspection and testing plan for all definable features of work. The inspection and testing plan defines the process for identifying how frequently and what items of work will be inspected and tested as required by the contract documents. It also specifies the appropriate self-inspections to control the work of the contractor and its subcontractors, suppliers, and fabricators to verify acceptance as defined by contract documents. Vertical construction has a similar process, which standardizes inspections for each jurisdiction and project stakeholder involved.

In addition to the project-specific inspection and testing performed to ensure the developer's interests, public inspectors and utility inspectors are also involved based on the local jurisdiction's standard practices and protocols. The contractor gives formal notice in advance to the appropriate inspection personnel to schedule project-specific inspections throughout construction. Public inspection departments have certified personnel that are knowledgeable in existing building, utility, and development regulations. They are qualified to carry out inspections of the facility and issue project permits for compliance with building codes, health and safety requirements, relevant plans, and construction specifications. Turning over public infrastructure such as roads or utilities to the public jurisdiction for acceptance into the public system requires documented public inspections. Permits and inspection departments will systematically examine all relevant building and site development plans to ensure that all construction details are in accordance with the applicable codes, standards, and guidelines.

Different types of inspection include:

Initial Inspection	The purpose of this procedure is to verify workmanship standards and dimensional requirements, review adequacy of control testing, confirm test results comply with the contract documents and approved submittals, and ensure the developer is in agreement with the quality of the work being performed.
Footings, Foundations, and Slab Inspections	Inspectors usually check the accuracy of installation of foundation supports, rebar, and excavated slabs, and positive drainage. Underground plumbing, mechanical, and electrical installations are also inspected.
Ongoing General or Miscellaneous Inspections	Continuous inspections conducted at various stages of the project. Inspections usually involve the testing of materials and/or supplies as they are installed and sometimes after they are in place. An example is erosion and sediment control compliance inspections and ongoing maintenance of these controls.
Offsite Inspections	Inspections to ensure that materials and products used are in accordance with the contract documents and the material suppliers are in adherence with their own QC procedures. This inspection usually includes a quality management program which provides periodic QA checks on high-risk suppliers and coordinates any applicable offsite testing.
Stop Work Notice Inspections	The stop work notice transfers decision-making responsibility to the appropriate level in the construction organization when major issues arise and typically occurs during routine inspections. Nonconforming products or processes can cause significant damage, rework, or violations to the quality plan, environmental plan, and/or the health and safety plan as determined by the Quality Program Manager, Safety Manager, and/or Environmental Manager. A stop work notice such as Fig. 6.6 is issued and tracked.
Final Inspection	Inspection takes place for each building trade upon completion of work and satisfaction of building permit requirements. After a facility's final inspection, a certificate of occupancy is issued, guaranteeing the safety of a building for occupancy. Certificates of occupancy are discussed in further detail in Chapter 8.

Stop Work Notice

Operation:	Date:
Location:	Segment:

Item of work:____________________ Foreman:____________________

Superintendent:____________________ Engineer:____________________

Description of Non-Conformance:

Due to the Non-Conformance to Project Specifications described above, this item of work shall be suspended.

Quality Program Manager ____________________ Date ____________________

Corrective action taken to resolve Non-Conformance

This item of work may resume at this time implementing the above mentioned corrective action.

Quality Program Manager: ____________________ Date: ____________________

Foreman: ____________________ Date:____________________

Form 1.16A –REV 00

FIGURE 6.6 Stop work notice form.

The FHWA *Construction Program Management and Inspection Guide* identifies several tasks associated with inspection and documentation as summarized in the following sections.

Preparation. Inspections are often a direct measure of compliance to the contract documents, such as the specifications and drawings. Knowledge of the specifications and the prescribed methods for measurement, adequacy and documentation is necessary prior to performing an inspection. During construction, the inspector also must understand the current scope, schedule, and any contract addenda associated with the work. A checklist with references to the project drawings or relevant specification sections can assist the inspection and review process, but checklists are not meant to limit the responsibilities of the inspector. An inspector should peripherally monitor all construction activities. Fig. 6.7 provides items to consider during a construction inspection.

INSPECTION COVERAGE	REVIEW OF WORK ITEMS

INSPECTION COVERAGE

1. Progress and quality of work
2. Materials and quality control
   a. Project sampling and testing program
   b. Project spatial provisions
   c. Computation and use of quality levels analysis
   d. Product acceptance
   e. Innovative materials
3. Workmanship
4. Construction operations and features
   a. Adequacy of provisions for safety and traffic handling (traffic management plan, traffic control plan, and public notification)
   b. Accelerated techniques
   c. Innovative processes and procedures
5. Project records
   a. Field checks by project personnel
   b. Quantity and quality of materials delivered, used, and rejected
   c. Construction work performed
   d. Methods and frequencies of checks on scales and other measuring devices
   e. Adequacy of field notes, diaries, and records supporting pay quantities
   f. Subcontracting
   g. Labor compliance, Equal Employment Opportunity, Disadvantaged Business Enterprise (DBE), and on-the-job training
6. Changes and extra work including time extensions
7. Compliance with environmental and American With Disabilities Act commitments and permit stipulations
8. Staffing, inspector qualifications, facilities, and project control
9. Claims and potential claims
10. Compliance with contract requirements with respect to physical measurements

REVIEW OF WORK ITEMS

1. Right-of-way clearance, demolition
2. Utilities
3. Clearing and grubbing
4. Earthwork and grading
5. Environmental
   a. Erosion and sediment control
   b. Dust abatement
   c. Construction noise
   d. Other environmental commitments
6. Storm sewer and structures
7. Work zone
   a. Traffic management and traffic control planning
   b. Installation and maintenance of traffic control devices
   c. Work safety
8. Paving
   a. Flexible
   b. Rigid
9. Site Furnishings
   a. Signs
   b. Lighting
   c. Fencing
10. DBE Performance
   a. Verify that DBE on the job matches DBE as proposed in bid
   b. Performing in accordance with contract commitments
   c. Directing its own activities
   d. Performing a commercially useful function
11. Miscellaneous
12. Cleanup

FIGURE 6.7 Construction Inspection Checklist (adapted from FHWA *Construction Program Management and Inspection Guide*, 2004).

Conducting the Review. The professionals that perform the inspection should be certified and knowledgeable in the subject matter. To verify that the work is in compliance with the drawings and specifications, the inspector should have an established set of procedures and documentation methods (see Fig. 6.8). The documentation should not include any generalized opinions and should

Specification Reference	Inspection Frequency	Objective	Action
101	Daily	Confirm the utility layout follows design documents	At time of survey, verify the proposed alignment
201	Daily	Verify environmental protection is installed and maintained	Cooridnate with contractor on current limits of work—compare work area with plans
301	End	Fence location should confirm to plans and specifications	Verify and document the location of fence, material, and installation/foundation stability

FIGURE 6.8 Sample inspection procedures.

focus on the facts and measureable observations. Comprehensive documentation that includes photographs, drawings, notes, and videos allows the inspector to backup any reports with evidence. Additionally, information collected from the field can be reviewed with other professionals to ensure the proper action is taken. Finally, the inspector should use the collected information to document findings.

Documentation. Documentation and reporting of inspection activities usually follows a prescribed format from a public agency or professional organization (see Fig. 6.9). The inspection report should organize the findings of the inspection activity. The narrative content should include the scope of the inspection, work progression, summary of findings, and necessary actions.

Nonconformance. All projects will need to have a nonconformance procedure in place before construction begins. A nonconformance is a discrepancy in characteristic, documentation, or procedure that affects form, fit, or function and thereby renders the quality of an item unacceptable or indeterminate in meeting the relevant project requirements. The purpose of the nonconformance procedure is to document the existence and location of nonconforming work, establish procedures for communicating the status of the issue, and verify the contractor addresses nonconforming work prior to installation. Common nonconformance items include placement of damaged materials or areas of poor workmanship that fail to meet the tolerances established in the project specifications and standards. Each project will have a specific flow chart for addressing nonconformance reporting and resolution dependent on all project stakeholders.

6.3.1 Building Commissioning

Building commissioning is the systematic process of verifying and assuring that building systems are designed, installed, carefully tested, and operational according to the developer's or building owner's specifications. It includes planning, inspection, and verification of a building's components after installation to ensure the building operates at optimal levels throughout its life cycle.

U.S. Department of Transportation
Federal Highway Administration

CONSTRUCTION INSPECTION REPORT

DIVISION	REPORT NO.	DATE OF INSPECTION	DATE OF REPORT	PROJECT NO.

INSPECTION MADE BY	QUALITY OF WORK ☐ Unsatisfactory ☐ Satisfactory	PROGRESS OF WORK ☐ Unsatisfactory ☐ Satisfactory	TIME ELAPSED %	WORK COMPLETED %

IN COMPANY WITH

(check appropriate box)
☐ Process Review/Product Evaluation ☐ Inspection-In-Depth ☐ Project ☐ Final

FIGURE 6.9 Sample inspection report form (FHWA Form 1446 A, rev. 10-89, from *Construction Program Management and Inspection Guide*, 2004).

Building commissioning ideally occurs prior to building occupancy and opening but can occur as early as the design development phase with a commissioning agent's review of design documents. Commissioning transforms the building system from a state of static completion to working order.

Commissioning includes setting-to-work, repetitive testing, and adjustments of the associated control systems. It also calibrates and documents the system settings. Testing of the building systems confirms the systems' performance meets the standards and specifications prior to the owner's acceptance of the building. Some building components such as the fire protection and life safety systems require commissioning as part of the final inspection and issuance of the certificate of occupancy. However, the developer may request additional building commissioning certifications for systems such as the audiovisual (AV) system. Building services requiring commissioning may include the following:

- Heating, cooling, and ventilation (HVAC) systems
- Electrical system
- Utility systems
- Building envelope
- Lighting
- Fire protection and life safety systems
- Elevators and escalators
- Audiovisual system
- Security systems
- Generators

Training for the building owner and/or representatives on equipment operation and maintenance should also be part of the building commissioning. Operations and maintenance is discussed further in Chapter 8.

CASE STUDY: SITE INSPECTIONS

ISSUE:

Design professional's liability for failure to note structural defect during site visits.

SUMMARY:

Architect designed lakeside vacation house and provided construction-phase services. A balcony 20 feet off ground was thoroughly designed with various sound structural features. A few years after completion, a pair of house guests stepped onto the balcony together and it collapsed, resulting in serious injuries. Forensic review revealed that contractor had failed to follow key design requirements, most notably by nailing the balcony to a thin sheet of plywood rather than bolting it to a 3-inch thick rim joist/blocking assembly as specified. The architect failed to notice defects, both in person and in photographs that established that the defects were open and obvious at the time of site visits and after completion. Jury assigned 90% responsibility to contractor and subcontractor and 10% to architect.

DECISION:

On appeal architect noted various standard AIA clauses that minimized the duty to detect construction errors, and asserted that the only duty was to report errors that architect had actually noticed. The court disagreed, holding that the architect had "a duty to identify observable, open, and obvious deviations that implicated safety and structural integrity and that were clearly presented to the architect."

COMMENTS:

The court distinguished the facts here from cases involving means and methods; from situations where defects are latent or obscured by other construction; where seeing the problem requires a particular vantage point; or involving insignificant deviations. The EJCDC construction-phase obligations attempt to reflect the balance that must be struck between the limited ability of an inspector to detect all problems and the owner's need for assurance that the construction conforms to the design. It is not contractually possible, or desirable from a policy standpoint, to create immunity for construction-phase A/E services.

Case study based on Black + Vernooy Architects v. Smith. Texas Court of Appeals (2010).

The information provided does not constitute legal advice. Consultation with an attorney is strongly advised.

6.4. CONCLUSION

The success of every construction project is anchored on profound construction management. Therefore, an experienced construction manager is knowledgeable in contracts, procedures, and documentation in order to navigate the intricacies of the project and best utilize project resources. Construction projects must comply with the standards of the applicable plans, specifications, codes, and other regulations. Accordingly, inspections and quality management will be carried out at various stages of the project to confirm the adherence to the contract specifications and regulatory codes.

The construction process is physically and intellectually demanding. All parameters and components crucial to achieve a cost-effective and timely project while satisfying the project's scope, safety, and quality requirements must be exhaustively evaluated and accomplished using best practices in every detail of the project.

CHAPTER 7

CONSTRUCTION TECHNIQUES

7.1. INTRODUCTION

The contractor is ultimately responsible for determining the sequencing of construction activities to meet the substantial completion date provided in the contract as well as implementing appropriate means and methods to construct the improvements detailed on the site plans. While civil engineers are not directly responsible for means and methods, they should consider means and methods relative to sequencing, cost, and constructability. Some site-specific conditions can be accounted for during design, but likely other unforeseen conditions will arise during construction that necessitate collaboration between the design team and contractor to resolve. A best practice is to perform constructability reviews with the contractor prior to construction commencement, which is inherently built into the construction manager at risk delivery method.

A site contractor typically installs the improvements shown on the site plans. The site contractor may contract directly with the developer for projects such as residential subdivisions or may be a subcontractor to the general contractor as is typically the case with commercial projects. In the latter case, the site contractor should coordinate scheduled activities with the general contractor, particularly on any site work items that impact the critical path schedule for the overall project and building.

7.2. TYPICAL SEQUENCING OF CONSTRUCTION ACTIVITIES

As discussed in Chapter 3, prior to the start of construction, the contractor will typically meet onsite with representatives from the local jurisdiction including inspectors to review the site and building plans, discuss procedures, and review the limits of disturbance. Prior to any land disturbance, the contractor will call Miss Utility to mark utilities within and surrounding the construction area. After utilities are marked, contractors will install perimeter erosion and sediment controls such as silt fence, tree protection, sediment traps or basins, and inlet protection on existing storm drains. Once these controls are in place, they will demolish any existing structures and/or pavement and clear any forested areas as indicated on the site plans. Depending on the size of the site, the demolition and clearing may be completed at the beginning of construction or in phases as the contractor works through the site. Once the required demolition and clearing has occurred, they typically begin striping topsoil, mass grading the site, and installing utilities. Based on the size of the project, they may also phase these activities. As discussed in the following sections, the type of development will dictate the critical path of site work activities such as installing pavement base course for laydown areas and access or grading the building pad. The storm drain, waterlines, and sanitary

FIGURE 7.1 Single-family residential development with ongoing infrastructure improvements. (Photo courtesy of William A. Hazel, Inc.)

sewer lines are typically installed beginning with the deepest utility (typically sanitary sewer). The contractor installs gravity lines such as storm drain and sanitary sewer downstream to upstream. Final site work activities include fine grading, installing pavement surface course and striping, landscaping, and installing other site furnishings to prevent damage from ongoing construction operations. Once the site is stabilized, the contractor removes any temporary erosion and sediment control measures such as silt fence and inlet protection.

7.2.1. Single-Family Residential

Single-family residential developments include residential subdivisions and townhome developments. Particularly with large residential subdivisions (also referred to as planned unit developments), a developer and homebuilder from separate companies are typically involved with the project. In this case, the developer installs the infrastructure improvements such as roads, utilities, storm drains, and stormwater management facilities and mass grades all the lots as shown in Fig. 7.1. The developer then sells the completed lots to a homebuilder to construct the home and ultimately sell to the homebuyer.

The sequencing of construction activities is based on the homebuilder's sales sequencing. For example, the homebuilder may request infrastructure improvements to be completed for the first 20 lots at the entrance of the neighborhood prior to starting foundations on any of these lots. These first lots often include the model home the homebuilder will set up as a sales office. The contractor will install utilities for the first phase, followed by installing curb and gutter and paving the roadway. While working on the paving for the first phase, the contractor will concurrently start installing utilities for the next phase. The phased infrastructure installation may align with the overall phasing plan for the subdivision, or the construction plans may include additional intermediate phases to accommodate the homebuilder's anticipated sales sequence.

After establishing perimeter erosion and sediment controls and completing any required demolition and/or clearing onsite, the contractor will install sanitary, storm, and water mains within or adjacent to the roadway. A sanitary sewer line may be installed dry, disconnected from the downstream public sewer main if the connection point to the public main is not within the initial work area. This prepares the initial lots for turnover to the homebuilder prior to installing the complete sanitary sewer system, thus reducing the time to complete the initial subdivision improvements. In this scenario, the contractor will connect to the public sanitary sewer during a later phase of infrastructure improvements and before any certificate of occupancy is issued. In contrast, per fire and building codes, the waterline typically has to be active with adequate fire hydrant coverage before vertical construction may begin. Therefore, the contractor may have to install additional waterlines beyond what is within the initial construction area to connect to the public water main.

Once the utilities in the area of the roadway are installed, the contractor will construct the roadway including the curb and gutter, subbase, and base asphalt layers. Typically the surface asphalt layer is not installed until the homebuilder is ready for the local jurisdiction to release the bond for the roadway improvements.

Once the infrastructure improvements are completed for the initial phase, the contractor will continue to work sequentially through the subdivision, installing the utilities, storm drain, and roadways in phases as determined by the homebuilder's sales plans. Once all the infrastructure improvements are complete, the contractor installs site furnishings such as benches and lights. Typically, the landscaping and surface course of asphalt are one of the final activities prior to project completion.

The sequencing for installing the stormwater management facilities will vary depending on what is proposed. A detention or retention basin will typically be excavated during the initial mass grading of the site and utilized as a sediment basin for erosion and sediment control throughout construction. Once the site is stabilized, the sediment basin is converted to the detention or retention basin. On the other hand, smaller, more localized stormwater management facilities distributed throughout the site such as bioretention or dry swales will be installed after other infrastructure improvements are complete but typically before the lot is turned over to the homebuilder. As these facilities require subsurface layers of stone and engineered soil material, they need to be protected with appropriate erosion and sediment controls to prevent sedimentation prior to final stabilization.

The duration for a large residential subdivision can take years from initial clearing to final completion and bond release for infrastructure improvements. The overall project schedule is largely dictated by the pace of the home sales.

7.2.2. Multifamily Residential

For multifamily residential developments including condominium and apartment complexes in suburban areas, the site contractor's critical path is to pave the proposed parking areas adjacent to the first buildings planned for construction in order to provide a laydown area for materials (see Fig. 7.2). Similar to single-family residential developments, the first buildings may be located closest to the site entrance and also include the leasing office and/or clubhouse facility. The site contractor will install utilities and storm drain within the proposed parking and laydown areas. The storm drain is typically connected to its outfall, so it can be used to drain the site during construction with appropriate erosion and sediment controls. Similar to single-family developments, the waterline and fire hydrants must be active during construction per the fire and building codes, but the sanitary sewer line may be installed dry if the initial buildings are not located near the sanitary sewer tie-in point. Based on the need to install utilities and storm drain initially, it is advantageous for the general contractor to start construction on buildings located closest to the utility and storm drain connection points. However, this may not be possible if the sales office and/or clubhouse are not located near utility connection points and are required to be the first buildings constructed.

Once the utilities and storm drain are installed adjacent to the initial buildings, the site contractor will install the parking lot curb and gutter, subbase, and base asphalt but typically reserve installing the

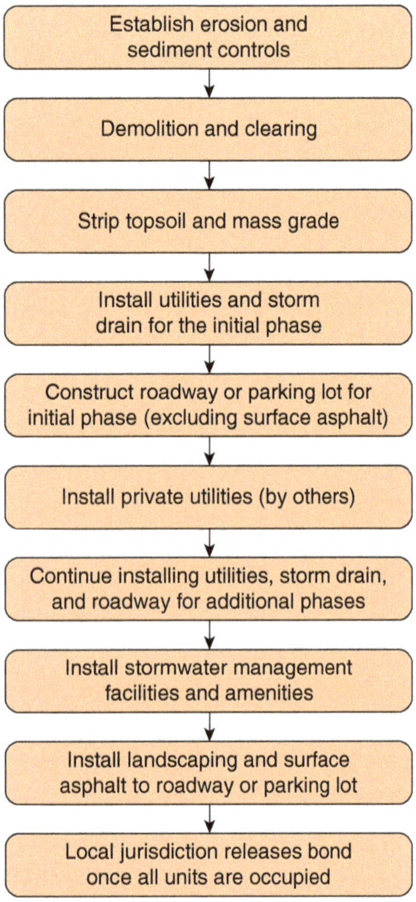

Figure 7.2 Typical sequence for single-family residential or multifamily residential projects.

surface asphalt until construction operations are close to completion. This approach minimizes the potential damage to the surface asphalt as a result of heavy equipment and use of these parking areas as laydown areas.

The site contractor continues installing utilities, storm drain, and parking areas throughout the site once the site work around the initial buildings is complete. Any stormwater management facilities, landscaping, and amenities onsite are also installed. Typically, all site work aside from the surface asphalt installation is complete before the buildings are complete.

7.2.3. Commercial and Industrial

For a commercial or industrial development, the critical path for the general contractor is vertical building construction. Therefore, the site contractor's priority is to prepare the building pad.

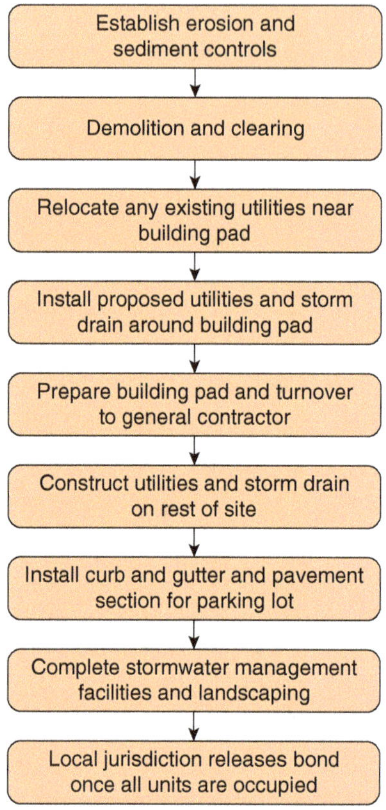

Figure 7.3 Typical sequence for commercial or industrial projects.

As shown in Fig. 7.3, the site contractor installs perimeter erosion and sediment controls and clears and grubs the site. Any existing utilities located near the proposed building pad will be relocated first. Any proposed utilities around the building pad will be installed concurrent with the mass grading of the overall site and grading of the building pad. The utilities around the building pad are installed prior to turning over the building pad to the general contractor to minimize impacts to utilities within the zone of influence of the structural building foundation.

Commercial developments in particular often utilize underground stormwater detention facilities for stormwater quality and quantity treatment (see Fig. 7.4). These facilities are typically located within parking areas, and depending on the size, may require a large excavation area to accommodate both the large diameter pipes and required bedding material. For this reason, the contractor often installs these facilities while mass grading the site and installing other utilities around the

Figure 7.4 Underground stormwater detention facility. (Photo courtesy of William A. Hazel, Inc.)

building pad. However, for smaller sites, space is limited for construction vehicle routes and laydown areas. If there will be frequent heavy construction equipment loading in the area of the proposed underground detention system, the contractor may choose to install the facility toward the end of construction when the building exterior is complete.

Once the building pad is ready for the contractor to begin vertical construction, the site contractor begins noncritical path activities including fine grading of the parking lot, pavement installation, grading stormwater management facilities, and installing landscaping. Public roadway improvements including turn lanes or sidewalks may also be performed during this stage. Site-work items are largely weather dependent, and depending on the schedule and time of year, the site contractor may demobilize or leave the site and remobilize or return closer to completion of the building to finish site work.

7.2.4. Special Sequencing Considerations

Certain site conditions or offsite constraints will influence the typical construction sequence on any project. During the design phase, the civil engineer should encourage the developer to use Miss Utility and/or a subsurface utility exploration (SUE) company to locate all existing utilities onsite as any required utility relocations can have a significant impact on the construction sequencing and schedule. Any existing utilities, storm drain, or streams traversing the site, particularly within the proposed building pad, will need to be relocated or rerouted immediately following the establishment of erosion and sediment control measures but prior to rough grading of the site. Any private utilities such as gas or electricity will require relocation by the utility provider, and including utility owner information in the topographic survey expedites the general contractor's coordination with the utility owner. Additionally, installing an initial phase of the storm drain and structures may be advantageous on an excessively flat site that does not drain well in the existing condition.

The local municipality may also limit the timing of any construction activities that disrupt public facilities. This may include disruptions to public water and sewer service or work within the right-of-way. Often, roadwork is scheduled for the middle of the day or during the summer to avoid peak traffic hours or in some cases, relegated to night work hours. The municipality may also limit work during certain events or seasons such as tourist seasons. This may impact not only the scheduling of roadway work but also louder construction activities such as driving piles for a foundation.

The contractor will also adjust the schedule to start activities requiring a long-lead time and coordination with outside organizations early in the construction period. This includes any relocation of private utilities such as utility poles, gas lines, or communication lines. As previously discussed in Chapter 5, the design of relocated private utilities typically does not begin until construction commences, and therefore, may require coordination over several months before the utility is relocated.

For projects on a campus setting including parks and schools, special consideration for maintaining access to buildings and site areas outside of the construction limits is required. This includes maintaining accessible routes, providing pedestrian detours, and providing appropriate fencing and signage around the construction area.

7.3. MEANS & METHODS

7.3.1. Erosion and Sediment Control

As indicated in the sequence of construction provided on the site plans, the contractor will install perimeter erosion and sediment controls as one of the first site work activities. Perimeter controls typically include silt fence, inlet protection on existing inlets, and tree protection. On wooded sites, the contractor will clear a path wide enough to install these controls but not disturb the rest of the site until all perimeter controls are in place.

After establishing the perimeter controls, the contractor will excavate the sediment traps or basins, if proposed, as well as any diversions or diversion dikes directing runoff to the traps or basins. Ideally, the civil engineer analyzes the sediment basin earthwork during the design phase and establishes a Phase I limits of disturbance that includes not only the installation of initial controls but also locations where the excavated sediment basin material is reused in proposed fill areas. Otherwise, the contractor will need to stockpile material, which requires moving the material at least twice. Less commonly, the contractor may coordinate with the public inspector to adjust the limits of disturbance, but often the inspector will uphold the site plan design.

Prior to mass grading the site, the contractor will strip the topsoil for use in landscaped areas or for export. The stockpile area should be within the limits of disturbance and in a location not conflicting with construction operations. If the stockpile area is not shown on the plans or interferes with construction activities, the contractor may coordinate a field adjustment with the public inspector. The stockpile will need to be mulched or seeded with the approved temporary seed mixture and surrounded by silt fence to prevent erosion during rainfall events (see Fig. 7.5).

The civil engineer should consider intermediate erosion and sediment control measures required

Figure 7.5 Stockpile surrounded by silt fence.

based on the construction sequencing. As an example, the contractor cannot construct a 30-foot fill slope in one day, and a single row of silt fence at the toe of the slope is inadequate to control erosion for the entire slope. Therefore, a row of silt fence perpendicular to the slope is required every 5 to 15 feet in elevation change to capture any sedimentation until the contractor can fully stabilize the slope. Once the fill slope construction is complete and heavy equipment is off the slope, the contractor can stabilize the slope with matting and seed, which may take several weeks to germinate. The contractor will add intermediate rows of silt fence perpendicular to the slope as the fill is placed. The contractor may add additional erosion and sediment control measures that are not identified in the site plans for temporary stockpile areas or to address additional requirements from the public inspector due to site conditions.

The contractor and public inspectors monitor erosion and sediment control devices on a regular schedule as well as after significant rainfall events. These devices require ongoing maintenance and replacement due to construction activities, wear and tear, and rainfall events. The local jurisdiction or state design standards provide maintenance specifications, and the site plans typically include a copy of the maintenance schedules. The contractor and the developer's representative (civil engineer, construction manager, or third-party inspector) maintains erosion and sediment control records including regular inspection reports.

7.3.2. Demolition

Building Demolition. Most local jurisdictions issue a demolition permit for any existing structures separate from the erosion and sediment control or grading permits, and therefore, selective demolition may occur prior to establishing the erosion and sediment control measures. In this case, the contractor will demolish the building but leave the foundation undisturbed as a stabilized surface until all perimeter erosion and sediment control measures are in place. Demolishing the foundation creates a disturbed area, thus requiring the erosion and sediment control permit and installed measures.

The local jurisdiction will typically require building utility disconnections and asbestos abatement, if needed, prior to issuing the demolition permit. The developer will coordinate with public and private utility companies to disconnect existing utilities to the structure. Additionally, any above or below ground fuel tanks will be removed by a specialty contractor. Drinking wells will be filled with slurry and capped under supervision of the Health Department, and old septic tanks are typically crushed in and backfilled. If the building contains asbestos materials such as flooring, roofing, or old boilers, an asbestos abatement contractor will remove these materials with appropriate personal protection gear and constant air monitoring. The removed asbestos material is sealed in bags, tagged, and disposed of at a facility that accepts hazardous waste. A local jurisdiction may have additional pre-demolition requirements beyond utility disconnects and asbestos abatement including rat-free inspections.

Existing concrete foundations are dug out and broken with an excavator-mounted hoe ram. The concrete material is crushed and recycled onsite so long as it is not contaminated. Any steel in the foundation is taken to a scrap yard.

Utility Demolition. Any existing pipes or culverts located within the proposed building footprint will be excavated and removed. Other pipes not within the building pad or in conflict with proposed site utilities will be abandoned in place if possible and filled with flowable fill. Flowable fill is less expensive and quicker than pipe removal. Larger diameter concrete or metal pipes may be removed even if they are not within the building footprint due to the large void space and potential for collapse, The excavated area will be backfilled with approved fill material. Backfill within the roadway is typically a stone material up to the asphalt layers, whereas compacted soil will typically be used in the proposed building pad, parking lot, or landscaped areas.

Ductile iron pipe will typically be taken to a scrap yard to be recycled. Due to the wire mesh, concrete pipe is difficult to recycle, and plastic pipe cannot be reused. If asbestos cement sewer pipe is present onsite, it will be taken to a certified landfill capable of handling hazardous material just like the asbestos material from the building.

Pavement Demolition. Pavement demolition can consist of full pavement removal or milling of the surface course. A loader or dozer will be used to remove the full pavement depth including the stone, base asphalt, and surface asphalt. A geotechnical engineer will evaluate the soil under the demolished pavement area. If the soil is deemed unsuitable and new pavement is proposed in this area, the contractor will undercut the pavement section and replace with borrow material to attain a stable subgrade.

A milling machine removes only the surface course for minimum grade adjustments or repair of damaged sections. Traditionally, milled asphalt was loaded into a dump truck to be transported to the asphalt plant as a recycled asphalt product (RAP). Alternatively, a scarifying machine mills the asphalt in place and incorporates it into the base stone or subgrade. The milled asphalt may also be used as a structural fill product with a geotechnical engineer's approval. This product yields an excellent structural fill material that is not as weather sensitive as soil and can minimize the amount of additional imported material required.

Clearing and Grubbing. Clearing and grubbing refers to the removal of trees or shrubs on a vegetated site, separate from stockpiling the existing topsoil. The cleared material is typically recycled as mulch (see Fig. 7.6) or firewood, or sold to a log mill. Similar to early building demolition, a contractor may be permitted to timber a site prior to installing the erosion and sediment control measures so long as the stumps are left in place. This as well as early building demolition should be coordinated with the local jurisdiction to understand their policies and procedures.

Endangered species and their habitat are of particular concern when clearing a site. Typically the bid documents will include information about potential species and/or habitats and restrictions on construction. This can include habitats and species located onsite or on adjacent sites. Certain habitats must be left undisturbed whereas other species may influence when clearing of a site may occur as to not interrupt nesting season. This may also influence the removal of site items such as light poles where birds nest. The developer may require these light poles to be removed before nesting season so new nests cannot be established and therefore disrupt the construction schedule.

Figure 7.6 Mulching. (Photo courtesy of William A. Hazel, Inc.)

The civil engineer should visit the site to review the clearing limits with the contractor and local municipality. The contractor typically avoids mobilizing the clearing subcontractor to the site multiple times, and therefore, remobilizing to adjust the limits of construction later may not be possible without a change order.

7.3.3. Dewatering

Dewatering is required for temporary stream diversions and any excavation below the groundwater table. To the extent possible, dewatering or stream diversions should occur during the drier months when the groundwater table is lower. For smaller, shorter duration stream diversions, a pump-around system may be used, but larger streams will require a berm to redirect the stream. Stream diversions tend to be labor intensive for both the construction of the initial diversion and maintenance.

Wellpointing and pumping are two options for dewatering. Wellpointing consists of multiple small diameter wells placed around the excavation site. A header pipe connects the wells to the pump, which creates a vacuum to remove the water from the excavation site. Wellpointing is more expensive but when installed correctly, is a faster, more reliable method to dewater an area. Alternatively, a pumping system may be used. The contractor can use the pump as needed to dewater an excavation site, and the pumped water will drain through a filtration bag (see Fig. 7.7) to prevent any silt from leaving the site.

FIGURE 7.7 Filtration bag for dewatering. (Photo courtesy of MBP)

7.3.4. Structural Components

Temporary Shoring. Temporary shoring may be required for pipe installation where the contractor does not have enough area to lay back or bench the soils per OSHA standards. Most commonly, a trench box is used to facilitate an excavation over 4 feet deep when the soils cannot be laid back. Trench boxes (see Fig. 7.8) are typically 20 to 24 feet in length and 6 to 8 feet tall. Trench boxes can be double stacked with a pinning system to yield additional depth for trench excavation at a depth of no more than 20 feet. The width of the trench boxes can be adjusted with different size spreaders.

Excavations for basements, underground parking areas, or foundations that are over 20 feet require a site-specific designed shoring system, sealed by a licensed civil engineer. This shoring is typically constructed with H piles and lagging where H beams are driven into the ground with $6 \times 6 \times 8$ timbers ("lagging") laid horizontally between the beams as the contractor excavates down. These sheet piles are sometimes removed but also can remain in place, cut off above finished grade. Shoring may also be required for deep or complex utility installation as shown in Fig. 7.9.

Foundations. For spread footings (see Fig. 7.10), the contractor digs a trench to match the footing dimension and then the rebar is placed on concrete bricks to serve as rebar chairs (see Fig. 7.11). The inspector will verify the subgrade is stable, and the rebar is installed per plan prior to the contractor pouring concrete. Often corner bars are mistakenly omitted where the spread footing turns 90 degrees at a wall corner and should be noted by the inspector as missing. Once the inspector has verified the subgrade and rebar, the foundation is poured (see Fig. 7.12).

Retaining Walls. Block retaining walls typically bear upon a concrete footer foundation, relying upon the weight of the wall and a stacked locking design to retain the earth behind them. For taller block retaining walls, a geogrid may be introduced between each course of block and extended into the backfill to provide stabilization. Block retaining walls are backfilled with an open graded aggregate and a corrugated drainage pipe to provide a means

FIGURE 7.8 Trench box for utility installation.

FIGURE 7.9 Temporary shoring for storm drain tie-in. (Photo courtesy of William A. Hazel, Inc.)

FIGURE 7.10 Marking for trench excavation for spread footing. (Photo courtesy of Fairfax County Park Authority)

Figure 7.11 Rebar chairs for spread footings. (Photo courtesy of Fairfax County Park Authority)

Figure 7.12 Completed spread footing foundation. (Photo courtesy of Fairfax Park Authority)

for water to drain from behind the retaining wall. Block retaining walls (see Fig. 7.13) are often used in residential or commercial areas because they are aesthetically pleasing.

Precast retaining walls are typically referred to as mechanically stabilized earth (MSE) retaining walls (see Fig. 7.14). MSE walls are often used in roadway construction or large earth retaining applications. The contractor first installs a concrete leveling pad for the precast panels to rest upon. The precast concrete panels interlock along their sides and have cast in steel anchors on the back. The contractor adds open-graded stone aggregate backfill with the panel installation. Steel straps connected to the panels are embedded into the aggregate backfill. The steel straps are typically longer near the base of the wall and shorter at the top of the wall. The geotechnical or structural engineer will specify the steel strap length based on soil conditions and load characteristics, but the strap at the base of the wall is typically about 70% of the wall height. Highly visible precast panels are often constructed at the plant with a form liner finish on the front face to make them aesthetically pleasing.

FIGURE 7.13 Segmental block retaining wall.

Site Furnishings. Site furnishings such as signage, light poles, and bollards typically have a subgrade foundation. While monument signs are typically site-specific, standard details in the site plans will often dictate the foundations required for the other site furnishings. As an example, parking lot lighting has a formed, cast-in-place foundation that protrudes approximately 2 to 3 feet above ground (see Fig. 7.15). Anchor bolts are placed on the top of the foundation to attach the light pole.

FIGURE 7.14 Mechanically stabilized earth (MSE) retaining wall.

Figure 7.15 Parking lot lighting foundation. (Photo courtesy of Fairfax County Park Authority)

7.3.5. Earthwork

Earthwork is typically one of the most expensive components of site work, and therefore, the developer and civil engineer should carefully review the specifications prior to bidding the project. With any earthwork operation, the contractor's goal is to move the soil once and avoiding stockpiling and relocating it later if possible. Prior to any mass grading or excavation, the contractor will strip the topsoil and stockpile it or immediately place it in nonstructural fill areas. Any stockpiled topsoil will be exported or spread in landscaped areas at final grade.

The contractor will perform a takeoff of the earthwork, determining how the project balances. The earthwork takeoff will include a cut and fill analysis as well as accounting for any spoils from pipe trenching, foundation footers, or retaining wall construction. The contractor will divide the earthwork into suitable ("good") versus unsuitable ("bad") soil, and where possible, any unsuitable soil will be placed in nonstructural areas. The contractor may build additional berms in landscaped areas to use unsuitable soils and exchange unsuitable soils for suitable soils from large greenspace areas. The worst-case scenario is having to export excess unsuitable soil and importing additional suitable soils to complete the project. Any export or import of soils is typically reserved as a later task so the contractor can confirm materials needed.

As mentioned in Chapter 3, contractors will bid the site as either classified or unclassified. For unclassified, they accept any soil conditions encountered between the existing grade and proposed subgrade shown on the site plans including the export of unsuitable soils and rock excavation. Contractors are responsible for excavating to the grade on the plans, and therefore, they assume the risk of determining the quantity and quality of excavated materials to the design elevation. For a classified site, the bidding documents will establish limits on the earthwork included in the base bid and provide a method for processing change orders such as unit price for any unclassified earthwork beyond the established limits. In this case, contractors are responsible for the quantity of excavated material between existing and proposed subgrade, but the developer assumes the risk of the quality of soil or rock. Separate from specifying the bid as classified or unclassified, the bid package will also state the contractor is responsible for a specified depth of undercut, such as 1 foot. Any undercut required beyond the specified depth necessitated by site conditions and directed by the geotechnical engineer is an additional cost to the developer.

Rock Excavation. The earthwork specifications combined with the boring logs prepared by the geotechnical engineer provide the contractor with a basis for the quantity of rock to expect onsite. However, if unexpected rock is encountered on a project with a classified earthwork bid, the developer will process a change order from the contractor for the additional work, likely on a time and materials basis.

Rock excavation can be completed with blasting, drilling, or hoeramming. Blasting with dynamite will cause swells in the rock material, up to 50%

of its original volume. This method is preferred as it is the least expensive and fastest. However, rock cannot be blasted close to existing utilities or buildings. In this scenario, hoeramming will utilize a large excavator with a hammer attachment to the boom to break up the rock. Perforating the rock with a drill will expedite hoeramming the rock.

Depending on the type of rock, the crushed material may be reusable onsite rather than requiring export. Rock can be generalized into three categories based on hardness, and each type creates challenges for site work. Diabase rock is the hardest rock, typically boulder-like, and hard to reuse. Hornfels rock is softer and more weathered and can be crushed with a sheepsfoot roller for use as structural fill with appropriate testing by the geotechnical engineer. Shale is one of the softest rocks that will break down after being stockpiled for approximately 30 days and can be placed in fill areas.

Placing Fill & Compaction. Any fill placed under buildings and pavement is considered structural fill and will need to be compacted to 95–100%. Fill in landscaped areas however may only need to be compacted to 85–90%. Depending on the size of the area and quantity of fill, the contractor will place fill with either an excavator or dozer and spread in 8-inch lifts. For large site development projects, the contractor will use an off-road truck to move material.

To construct a building pad in fill, any soft or yielding subgrade will first be undercut under the monitoring and certification of a geotechnical engineer. If the subgrade yields due to excess moisture, the contractor can lime the subgrade if the material is otherwise suitable. After each lift, a compaction test will be performed, and the geotechnical engineer will prepare a density test and special inspections report. Many projects will require the top layers of fill to have a higher California Bearing Ratio (CBR). Therefore, when multiple lifts are required to achieve the proposed finished grade, lesser quality structural fill will be placed at the bottom with the first lifts, and the highest quality fill material will be used at the top for the last lifts.

To construct a building pad in cut, the geotechnical engineer will certify the bearing capacity of the subgrade. If it is not suitable, undercut and structural fill is required to achieve the subgrade elevation.

Compaction equipment will vary based on the application and area. Compaction in small trenches will typically be done with a jumping jack tamper, and larger trenches will use a trench roller. Large fill areas will use a smooth drum or vibratory sheepsfoot roller until near the subgrade or finished floor elevation. A motor grader or dozer will fine grade the subgrade or building pad and adjust the grade to final elevation.

Undercut. Undercut is when unsuitable soil conditions are encountered in areas where structural improvements are proposed including buildings, retaining walls, roadways, and parking lots. Based on the subsurface investigation completed during the design phase, the geotechnical engineer may recommend a depth of undercut in certain areas based on the soil borings and proposed improvements. Depending on the scope of work, the contractor may include a line item for undercut in his bid.

However, soil borings provide a limited sample of the overall subsurface conditions of a site, and it is possible additional undercut or removal of unsuitable soils will be required once the contractor begins grading the site. The special inspector will provide oversight on the undercut required based on construction conditions. Additionally, exposed soils can be impacted by weather conditions, specifically rain or frost. Grading the site under wet conditions reduces the strength of clay soils in particular and likely will require additional undercut. If the schedule allows, the saturated soils may be excavated, stockpiled with moisture conditioning to dry out, and reused onsite as structural or nonstructural fill. Structural soils that are wet may be laid out to dry and scarified daily, though the soil is subject to additional rain prior to being placed in fill areas. During wet seasons, a contractor may lime the soils to expedite the drying, which can be expensive. Handling wet soil in either manner will increase the construction schedule.

Geo-Fabrics and Stone. In areas where unsuitable soil is anticipated beneath structural fill, the site plans may indicate additional layers of angular stone

Figure 7.16 Installation of concrete storm pipe. (Photo courtesy of MBP)

are required under the slab. This creates a stable base that is self-compacting, and the contractor will use a tamper to lock the stone in place.

For roadways, a geotextile fabric may be required on top of the subgrade to bridge unsuitable soils or to offer additional reinforcement between the subgrade and base stone layers. This fabric may be helpful during construction to provide a more stabilized access road that supports frequent, heavy equipment loading.

7.3.6. Utility Installation

The contractor will install the deepest utility first, which is typically the sanitary sewer but varies based on site conditions. When utilities such as waterlines and storm drains are at approximately the same depth, the contractor may alternate between installing each utility. The trench for the pipe is typically dug with an excavator, and the trench width is approximately double the size of the pipe. The base of the trench is inspected for stability and the presence of water. The pipe bedding requirements will be provided in the site plan and specifications. Typically, a commercially graded stone is required for the bedding material underneath the pipe (see Figs. 7.16 and 7.17). Smaller diameter plastic pipes such as roof drains may require pea gravel as the bedding material. For storm drain, concrete pipes are often backfilled with onsite soil above the spring line of the pipe as they do not deflect whereas a plastic pipe will be encased with stone on the sides and above due to the potential for deflection under compression. Because of the additional bedding material requirement, plastic storm drain installation tends to be more expensive than the equivalent diameter concrete pipe. Once the bedding material is installed per the plans and specifications, onsite soils are placed in 8-inch lifts and compacted on top of the pipe.

The local jurisdiction may require a camera inspection of installed utilities. The video confirms the deflection of the pipe is within the tolerance specified by the jurisdiction's construction specifications. While smaller diameter plastic pipe used for water or sanitary sewer applications deflects less, larger diameter plastic storm drains experience greater deflection, which is why concrete may be preferred for both cost and constructability.

FIGURE 7.17 Waterlines connecting to a proposed building. (Photo courtesy of William A. Hazel, Inc.)

Connecting to Existing Utilities. Connections to existing waterlines are typically done as wet taps with a tapping sleeve and valve to not disrupt water service for adjacent sites. The tapping sleeve is a tee split in two pieces installed on either side of the existing waterline, and the two pieces are bolted together. The valve is located on the branch line that connects to the tapping sleeve (see Fig. 7.18). With the valve open, an airtight tapping machine is connected to the branch line and drills a hole in the side of the existing waterline. As the tapping machine is pulled out of the branch line, the valve is closed to prevent any leakage.

Alternatively, a section of the existing waterline may be closed off to allow the installation of a new tee. The contractor will close existing valves on the existing waterline on either side of the proposed tee and drain the water from this waterline section. A cut-in-tee will be installed at the connection point. This method is cheaper but also requires disruption to utility service, which may be undesirable for the local jurisdiction.

For sanitary sewer connections, the invert of the proposed sewer line is ideally above the crown of the existing sewer main, and the contractor can

FIGURE 7.18 Tapping sleeve and valve installation. (Photo courtesy of Loudoun Water)

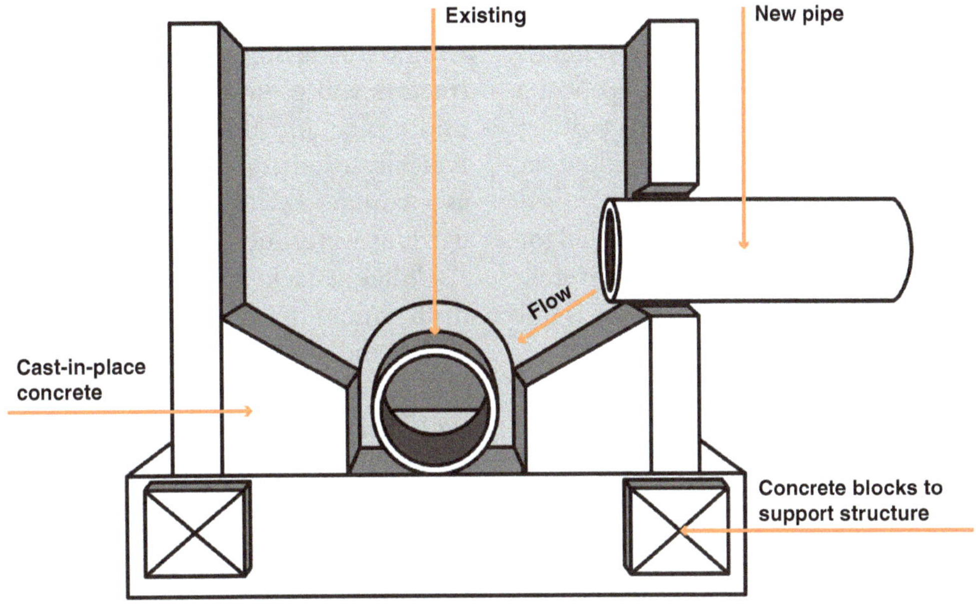

FIGURE 7.19 Doghouse manhole.

install a doghouse manhole. With a doghouse manhole, the contractor will pour a footing under the existing pipe and add the doghouse manhole (see Fig. 7.19) on top with a connection for the proposed sewer line. The contractor will then cut off the top of the existing sewer pipe to allow the proposed sewer line to drain into the existing system.

Where a doghouse manhole is not allowed either based on design constraints or the jurisdiction's standards, an external bypass is required. As shown in Fig. 7.20, the contractor adds a bulkhead to the upstream manhole (C) to block off the invert out and places a pump in the bottom of the manhole. The pump bypasses the sanitary

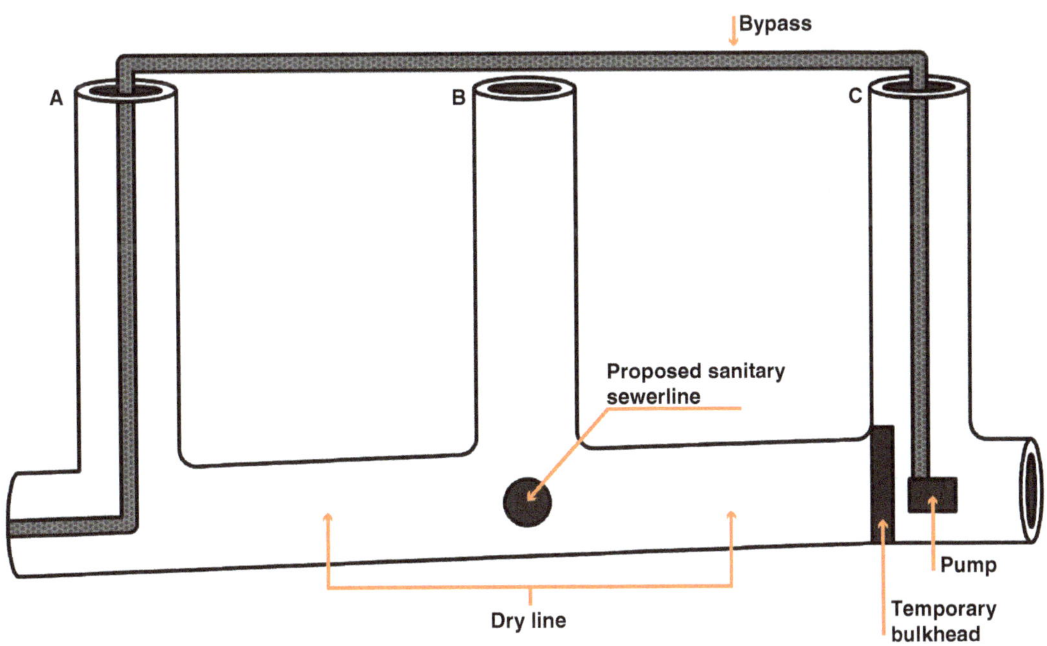

FIGURE 7.20 Sanitary sewer bypass.

sewer to the downstream manhole (A). Manhole B is the location of the proposed sanitary sewer connection and can either be an existing or proposed manhole. By using the bypass, the sanitary sewer line between Manhole A and C is dry to allow for the proposed connection.

For storm drain, the connection is scheduled for when the pipes are dry outside of rainfall events, and therefore a doghouse manhole or bypass is not required.

Trenchless Utility Installation. Sometimes contractors cannot excavate a trench for a utility installation such as open cutting an existing roadway or installing a utility under a waterway. In these situations, they may use a trenchless utility installation method such as jack and bore or horizontal directional drilling. For jack and bore, contractors will excavate a jacking pit (typically 20' × 40') and a receiving pit (typically 12' × 12' to 20' × 20') on either side of the roadway or waterway. The boring machine is placed in the jacking pit (see Fig. 7.21), and the machine's auger creates a tunnel beneath the surface, pushing the casing pipe through. The utility pipe will be installed within the casing pipe without disturbing the surface. In some cases, contractors will grout the annular space between the utility pipe and the casing pipe. Jack and bore is typically used for rigid pipe laid on-grade, such as a sanitary sewer or storm drain but can also be used for waterline or large ductbank installations.

Unlike a jack and bore, horizontal directional drilling does not require pits and instead starts and ends at the earth's surface. The drilling machine will create a tunnel under the surface and can adjust to avoid existing utilities. An initial, smaller diameter drill will be used first, followed by a larger diameter drill which is larger than the proposed pipe diameter. The pipe is pulled through the tunnel, and because of the variable path to avoid existing utilities, flexible, smaller diameter pipe is preferable for directional drilling applications.

7.3.7. Stormwater Management

Frequently, stormwater regulations require low-impact development (LID) with best management practices (BMP) that promote infiltration

Figure 7.21 Jacking pit for jack and bore operation. (Photo courtesy of William A. Hazel, Inc.)

of stormwater runoff and groundwater recharge instead of the traditional retention ponds. Examples of BMPs include bioretention (or rain gardens) and dry swales. These facilities require subsurface layers of stone (see Fig. 7.22) and engineered soil material (see Fig. 7.23) to promote infiltration as well as potential underdrains connecting to the storm drain system. While these facilities are typically shallower than a traditional retention pond, the contractor will typically prefer to excavate these areas while completing the mass grading of the site when the heavy site equipment is mobilized onsite. The risk with installing these facilities early in the project is that they may be contaminated with site soils during heavy rains. If the BMP is excavated early in the project, it is preferable for the stone, soil layers, and plantings to be installed at the very end of the project to minimize contamination. Additional erosion and sediment control measures including super silt fence and diversion dikes should surround any BMP constructed prior to site stabilization, and the base should be lined with a filter fabric.

These BMPs often require plantings at the base which can be trees, shrubs, or grasses. Grasses, specifically plug plantings, may be preferred as seed mixes may take multiple years to establish, tree roots can compromise the underdrains, and shrubs may not be maintained.

The pipe underdrain is typically designed at a flat slope due to design regulations; however this is not constructible. Where possible, the underdrain slope should be increased to 1–2% to promote better drainage. Additionally, the underdrain should not be wrapped in filter fabric. The filter fabric is only placed between the gravel and engineered soil media sections.

7.3.8. Pavement

After mass grading the site and installing storm inlets, the contractor will complete fine grading necessary to meet edge of pavement grades indicated with the construction stakeout survey. Pavement installation will begin with curb and gutter to create a border for the asphalt or concrete pavement installation (see Fig. 7.24). Concrete pavement can

Figure 7.22 Stone section for bioretention. (Photo courtesy of Fairfax County Park Authority)

Figure 7.23 Engineered soil media for bioretention. (Photo courtesy of Fairfax County Park Authority)

Figure 7.24 Parking lot with curb installed prior to asphalt installation. (Photo courtesy of Fairfax County Park Authority)

also use forms for an edge. Prior to installing pavement or curb and gutter, the contractor should verify the tie-in elevations shown on the site plans with field conditions and coordinate any discrepancies with the civil engineer.

Curb. Straight sections of curb and some radial curb will be constructed with a curb machine, which utilizes a string line. Curb and gutter installed with the curb machine needs to have a standard section such as 1":1' catch or 2% spill gutter pan. Gutter pan with a variable cross slope cannot be installed with a curb machine and is installed by-hand. As shown in Fig. 7.25, pins are set behind the proposed curb location with a string connecting the pins. The elevation of the string varies with the curb elevation. For radial curb, pins are set more frequently, approximating the curve radius with the chord length. The curb machine works best on radial curves with a radius greater than 30 feet; however, it can be used with radii as small as 5 feet around parking lot islands. Some curb machines utilize global positioning system (GPS) modeling which does not require a string line around curves.

The contractor will install some curb by hand, which is less efficient than the curb machine. This includes radial curve with small radii and gutter pans with variable slopes transitioning between catch and spill gutter. The contractor will use forms to set up the curb. The elevations of the gutter pan flowline should be confirmed prior to construction to confirm positive slope through any transition sections.

Site plan details and specifications will include requirements for stone layers and an underdrain beneath the curb.

Asphalt. Asphalt, also known as flexible pavement or asphaltic concrete, consists of multiple layers with variable thickness depending on the anticipated traffic loading. The basic layers of asphalt include a prepared subgrade, aggregate layer, and asphalt layer. The contractor will install the aggregate base on top of the prepared subgrade (see Fig. 7.26), and for heavier loading areas, the aggregate base may include a base and subbase layer as indicated in the geotechnical report and site plans. The contractor may use cement treated aggregate (CTA) as an alternate to the traditional

FIGURE 7.25 Installation of curb and gutter prior to asphalt pavement. (Photo courtesy of MBP)

FIGURE 7.26 Installation of base aggregate in background, compacted subgrade in foreground. (Photo courtesy of William A. Hazel, Inc.)

base aggregate. CTA has a stronger strength value than base aggregate and therefore, with the geotechnical engineer's approval, requires less aggregate depth for the pavement section, decreasing the overall cost. CTA may also be specified in the site plans for heavy duty pavement. Additionally, CTA is less susceptible to weather allowing for quicker paving during wet seasons. With CTA, contractors can typically pave the day after a rain event, whereas with standard base aggregate, they will have to scarify the stone for a couple of days to dry out. The aggregate base layer will be compacted, and a prime coat will be applied between the aggregate and asphalt layers (see Fig. 7.27). As specified in the geotechnical report, a typical pavement section for local streets requires 6-inch subbase aggregate, 4-inch base asphalt, 3-inch intermediate asphalt, and 1.5-inch surface asphalt. Interstate highway pavement sections are thicker, whereas parking lot pavement sections are thinner. For low traffic areas such as parking stalls, only a surface asphalt layer of 1.5 inches to 2.5 inches will be installed on top of the aggregate, and higher traffic areas within parking lots such as the truck route will have both an intermediate and surface layer of asphalt added on top of the aggregate. After placement by the asphalt paver, each asphalt layer will typically be rolled with a tandem drum roller. Asphalt installation is weather dependent and depending on the jurisdiction's or Department of Transportation's (DOT) standards, typically requires a minimum ground temperature of 40 to 50 degrees and rising to install.

Prior to installing asphalt, elevations should be double checked to ensure proper drainage. Ideally a parking lot is graded with straight or radial grades over larger areas. Additionally, consistent longitudinal and cross slopes increase constructability. Varying the design slope over short distances is not constructible with an asphalt paver, which installs asphalt in sections approximately 10 feet wide. Additionally, it is difficult for the paving machine to create ridges or swales in areas within 10 feet of the curb due to the asphalt paver width.

For asphalt parking lots, it is common for the civil engineer to propose a different pavement section

Figure 7.27 Installation of base asphalt layer on top of aggregate layer. (Photo courtesy of MBP)

for drive aisles versus parking stalls. Where possible, the asphalt surface layer depth should be consistent between the two pavement sections for constructability and future maintenance such as mill and overlay. As an example, the drive aisle pavement section is 6 inches aggregate subbase, 2.5 inches aggregate base, and 1.5 inches surface asphalt, and the adjacent parking stall pavement section is 8 inches aggregate subbase and 2 inches surface asphalt. This creates a 0.5 inch discrepancy in the top elevation of the aggregate section which is challenging for contractors to maintain and also prevents them from installing a consistent asphalt surface layer depth across the entire parking area.

As discussed previously with pavement demolition, if the existing subgrade and aggregate layers of a roadway are acceptable, the civil engineer may propose a mill and overlay of the existing surface asphalt layer to modify existing roadway grades. Raising the edge of pavement grade by increasing the asphalt surface layer depth is more constructible than decreasing the elevation. If the edge of pavement needs to be lowered significantly, the existing asphalt surface layer in addition to a portion of the intermediate asphalt layer and/or aggregate layer may require removal, which compromises the strength of the pavement section. In this case, the contractor may reuse milled asphalt or structural fill material to mix with the existing base stone and subgrade to increase the aggregate depth.

As with new asphalt pavement sections, elevations in the mill and overlay area should be double checked, particularly the existing roadway centerline elevation versus the proposed edge of pavement to ensure an adequate cross slope is maintained. Typically with a mill and overlay, the milling machine will remove 1.5 inches to 2 inches of the surface asphalt layer, and the asphalt paver will relay an asphalt surface layer of at least the same depth as the removed section. An overlay greater than 6 inches will likely require a multiple-day lane closure and also may be more expensive than installing a new pavement section. Mill and overlay is also used for maintenance of existing roadways and parking lots (see Fig. 7.28).

FIGURE 7.28 Mill and overlay of existing drive aisle. Milled pavement shown on right side with overlay installed on the left. (Photo courtesy of Fairfax County Park Authority)

Concrete. Concrete, also known as rigid pavement, consists of an aggregate layer and a concrete slab. As with asphalt pavement, the contractor will install the aggregate layer on top of the prepared subgrade, compact the aggregate, and top with the concrete slab. The aggregate layer and concrete slab thickness will vary based on anticipated traffic or nontraffic loading. Some concrete sections may also specify a reinforced welded wire fabric (WWF) which is placed on top of the aggregate before the concrete slab is poured. Concrete is highly susceptible to cracking, and control joints must be strategically placed to control cracking. Oftentimes, dowels are introduced at expansion joints to allow the concrete pavement to expand and contract without differential settlement occurring at the joint. Concrete is typically used in service or loading areas due to the heavy vehicle loading from delivery trucks and trash trucks. Concrete should also be specified in any maneuvering areas for tractor-trailers as concrete is more durable and not susceptible to the same rutting as asphalt pavement. Sometimes concrete is also specified in accessible (ADA) areas including accessible parking areas. Concrete has a lower construction tolerance than asphalt pavement and is therefore preferred in areas with critical or flat slopes. As an example, an accessible trail made with concrete may be designed up to the accessible longitudinal and cross slope limits of 5% and 2% (1/4" per 1') respectively, whereas an asphalt trail will need to be designed with maximum slopes of 4.5% and 1.5% to allow for variability in construction.

Concrete uses forms (see Fig. 7.29) or installed curb for its edges. As an example, for concrete sidewalks placed in landscaped areas, the forms will be set up on top of the aggregate before the slab is poured. Once the concrete is poured, floating will create a smooth concrete surface (see Fig. 7.30).

Concrete is also not subject to the same temperature constraints as asphalt. Additives included in the concrete mixture can aid in cold-weather placement, and typically concrete can be installed at above-freezing temperatures. If freezing temperatures are

Figure 7.29 Concrete sidewalk installation. (Photo courtesy of MBP)

expected overnight while the concrete is still curing, blankets can be used to cover the concrete.

As with asphalt pavement, elevations should be double checked prior to installation of the concrete pavement. In particular, formwork for any sidewalks or curb ramps should be checked for conformance with accessible slopes rather than checking slopes after the concrete is poured.

Figure 7.30 Floating concrete to provide smooth surface.

CASE STUDY: CONCEALMENT

ISSUE:

Effect of architect's concealment of cause of defect on the statute of limitations.

SUMMARY:

The architect's design of a community center called for removal of 6 feet of unsuitable peat soil. The contractor did not excavate to that depth, but the architect agreed that the excavation was adequate and authorized the deviation.

The hole was then backfilled. After the project was completed, the building's concrete floor, which was floating on the soils and backfill below, began to settle. The Town delayed paying the contractor, who eventually sued the Town for full payment. During the lawsuit, some five years after completion, the Town named the architect as a third-party defendant, based on the architect's approval of the deviation from the design requirements for excavation.

The applicable statute of limitations was four years. A standard AIA clause indicated that by mutual agreement, the statute of limitations would start running no later than the date of substantial completion. This clause effectively eliminated any discovery rule from being applied to extend the statute of limitations. The architect moved for summary judgment based on the expiration of the four-year limitations period.

The Town countered that the architect had concealed and misrepresented the cause of the building defect, which resulted in the Town not asserting its rights until the facts finally became known; therefore, the architect should not be allowed to be dismissed from the case. The architect had repeatedly told the Town that the problem was the contractor's failure to compact the backfill, and the contractor's related failure to conduct subgrade soil compaction testing during construction.

The trial court ruled in favor of the architect, and the Town appealed.

DECISION:

The appellate court agreed that a contract clause can eliminate the discovery rule from a statute of limitations. Thus, the applicable statute of limitations was four years from substantial completion. However, the court ruled that equitable estoppel based on concealment could affect the viability of a late claim. The critical question was whether the Town had sufficient information to pursue a claim within the four-year period. The appellate court held that it did.

The Town knew that the architect had authorized the deviation from the excavation design. The Town also knew that the floor problem was a direct result of poor underlying soils (as opposed to the backfill), based on an expert report two years after completion. The court held that the Town should have been able to put these facts together to reach the conclusion that the architect (via its approval of the deviation) might be partly to blame for the problem. As to the architect's repeated attempts to shift blame to the contractor, the court held that the Town could not turn a "blind eye" to the facts.

The Town also contended that the architect should be liable because of a fiduciary relationship to the Town—a position of trust, confidence, influence, and superiority. The court held that the two parties were operating at arm's length pursuant to a formal contractual agreement. This would not suggest a fiduciary relationship.

COMMENT:

EJCDC's professional services agreements contain clauses that start the relevant limitation period at no later than substantial completion; this case expands the list of cases that have supported the enforceability of such clauses.

The situation described here between architect and owner is very common. The design professional often has a close working relationship with the owner, and may have earned the owner's respect and approval over a period of years. Nonetheless, when a serious problem occurs, the prudent owner will need to be objective and explore the possibility that the design or other A/E services are at least partly involved—and will do so well before the right to take action lapses.

Case study based on J.S. Riemer, Inc. v. Village of Orland Hills. Appellate Court of Illinois, First District (2013).

The information provided does not constitute legal advice. Consultation with an attorney is strongly advised.

7.4. CONCLUSION

The final decision for construction means, methods, and sequencing rests with the contractor. However, civil engineers should consider construction techniques during design development. They can build their knowledge of these techniques and approaches through constructability reviews and collaboration with the contractor on every project.

When construction issues arise, civil engineers should always gather information on what improvements are installed or will be affected by any design changes and also understand any site condition constraints that will drive the most cost-effective and efficient solution. These lessons learned and experiences should be applied to future projects.

CHAPTER 8

CONSTRUCTION CLOSEOUT

8.1. INTRODUCTION

As a project nears completion, developers seek assurance that the improvements are constructed in accordance with the construction documents. The civil engineer's role in the project closeout process includes conducting site visits to evaluate substantially completed site work, reviewing as-built drawings and surveys against the construction documents, and drafting certification statements for specific improvements. The developer's ultimate goal is to obtain a certificate of occupancy and maintain a set of documents that allow for efficient operations and maintenance (O&M) as the facility and/or infrastructure ages. Additionally, the documentation produced by the civil engineer during project closeout is often required by the local jurisdictional for any public site improvements prior to issuing a certificate of occupancy. The certificate of occupancy is often the final permit developers need in order to begin using the facility and generating a return on their investment.

8.2. PUNCH LIST

A punch list details incomplete or unsatisfactory improvements that should be addressed by the contractor prior to substantial completion. A project may have multiple punch lists prepared by different parties. The local jurisdiction will often prepare a punch list for public improvements including utilities, roadway, stormwater, and landscaping as part of the final inspection. The contractor will address any deficient work items prior to the issuance of the certificate of occupancy and/or release of the performance bond. Examples of items listed on the local jurisdiction's punch list include the following:

- Leaking connection at water meter
- Adjusting manhole lids or valve boxes to finished grade
- Sidewalk slopes exceeding accessibility requirements
- Inadequate outlet protection
- Dredging accumulated construction sediment from stormwater facility
- Inadequate caliper of street trees

Additionally, the developer may have civil engineers and architects prepare a punch list for site work and building improvements respectively. Civil engineers should note any discrepancies between the construction documents and completed improvements as well as any deficiencies. They should also provide accompanying photos for any improvements needing attention. The full punch list should be provided to the developer and contractor, and it is at the developer's discretion which items to require the contractor to complete in order to release the retainage from the contract. Examples of items listed on the civil engineer's punch list include the following:

- Cracking pavement
- Poorly graded pavement that retains stormwater
- Leaking irrigation systems
- Missing landscape and inadequate depth of mulch
- Missing site furnishings including signs and wheel stops
- Inadequate permanent stabilization of pervious areas
- Removal of temporary erosion and sediment control measures

An example of a developer's punch list for onsite improvements is included in Fig. 8.1.

After the contractor has addressed the punch list items, the public inspector and/or the civil engineer may return to the site to confirm.

8.3. AS-BUILT AND RECORD DRAWINGS

As-built and record drawings reflect the modifications that occur to the construction drawings as a result of field changes, requests for information (RFIs), and developer-directed plan revisions. They are provided to the developer and local jurisdiction for their records. These drawings provide the local jurisdiction with a final opportunity to verify that the construction complies with local regulations and standards. Furthermore, developers, occupants, and management organizations use the as-built information for O&M purposes. The contractor or public inspector prepares as-built drawings, typically as red lines on the construction drawings or as a separate sketch (see Fig. 8.2). Additionally, the local jurisdiction or the developer may require an as-built survey for certain site systems such as the storm sewer or sanitary sewer network. The as-built survey will include horizontal location as well as rim and invert elevations and is prepared by the construction surveyor. The contractor and construction surveyor are responsible for the accuracy of the as-built drawings and survey respectively. Depending on the local jurisdiction's requirements, the as-built survey may include the following site work information:

Washington County
Parks and Recreation Department

February 1, 2021

ABC Construction Company
Attn: John Smith
456 Orange Avenue
Washington County, CA

Dear Mr. Smith:

The substantial completion inspection for the Main Street Recreation Center was completed on January 27, 2021. Attendees for the inspection were as follows:

- Arnold Watts, Washington County Parks and Recreation
- Melissa Jones, Washington County Engineering
- Steve James, Washington County Planning
- Suzie Jones, Dewberry

The following deficiencies (associated photographs provided in Attachment A) were noted as requiring correction or completion before final acceptance of the project:

General

1. Submit site as-built drawings to Washington County Engineering for review and approval.
2. Submit all special inspection reports.
3. Provide O&M manuals to the Parks and Recreation department.

Site Concrete

1. Replace cracked concrete in ADA ramp adjacent to building's north entrance.
2. Sidewalk south of building exceeds 2% cross slope required for accessibility.

Parking Lot

1. Stripe access aisle between accessible spaces.
2. Address water ponding at east site entrance.

Landscaping

1. Mulch layer to be 3".
2. Apply additional seed and matting to denuded areas.

This list of specific items is intended as a guide to ABC Construction Company to use in completing your work in an acceptable manner. Other items not in compliance with the contract documents or applicable codes are required to be corrected even if not noted above.

Sincerely,
Arnold Watts
Project Manager

Attachments: A – Representative Photographs

FIGURE 8.1 Developer's punch list.

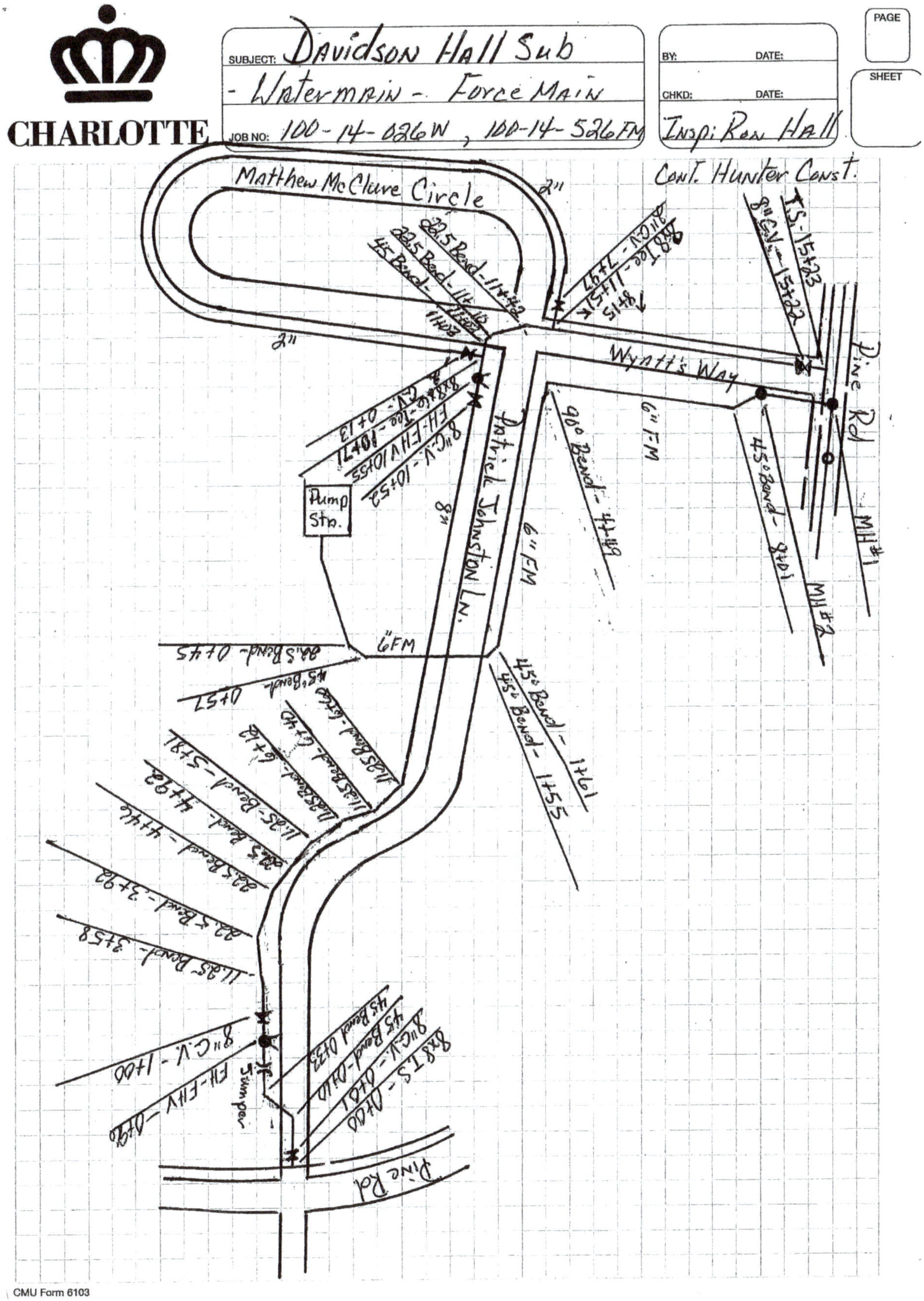

FIGURE 8.2 Inspector's as-built notes for waterline.

1. **Water**—Location of all water utilities constructed including dimensions to houses and structures. This allows the water utility provider to locate the utilities in the future even under adverse weather conditions and recover them in the event the utility is damaged. In many local jurisdictions, the fire hydrants must be located in the North American Datum (NAD 83) so they can be input into the emergency 911 system for reference during an emergency.
2. **Sanitary Sewer**—Location of all sanitary sewer utilities constructed including pipe and structure size, type, and slope (see Fig. 8.3). Comparison of this information to the approved plans allows confirmation of adequate conveyance systems. This also guarantees all sewer utilities are located within the right-of-way and/or easements.
3. **Storm Sewer**—Location of all storm sewer pipes and structures including size, type, and slope. Similar to sanitary sewer, comparison of the built condition to the approved plans allows confirmation of adequate conveyance systems. This also guarantees all storm sewer pipes and structures are located within the right-of-way and/or easements.
4. **Stormwater Management Facilities**—Location and size of the stormwater management facility including associated pipes and structures such as outlet control structures and emergency spillways. This as-built survey ensures the stormwater management facility provides the correct volume of stormwater per the approved site plans and calculations. This also guarantees the stormwater management facility is located within an easement or designated lot.

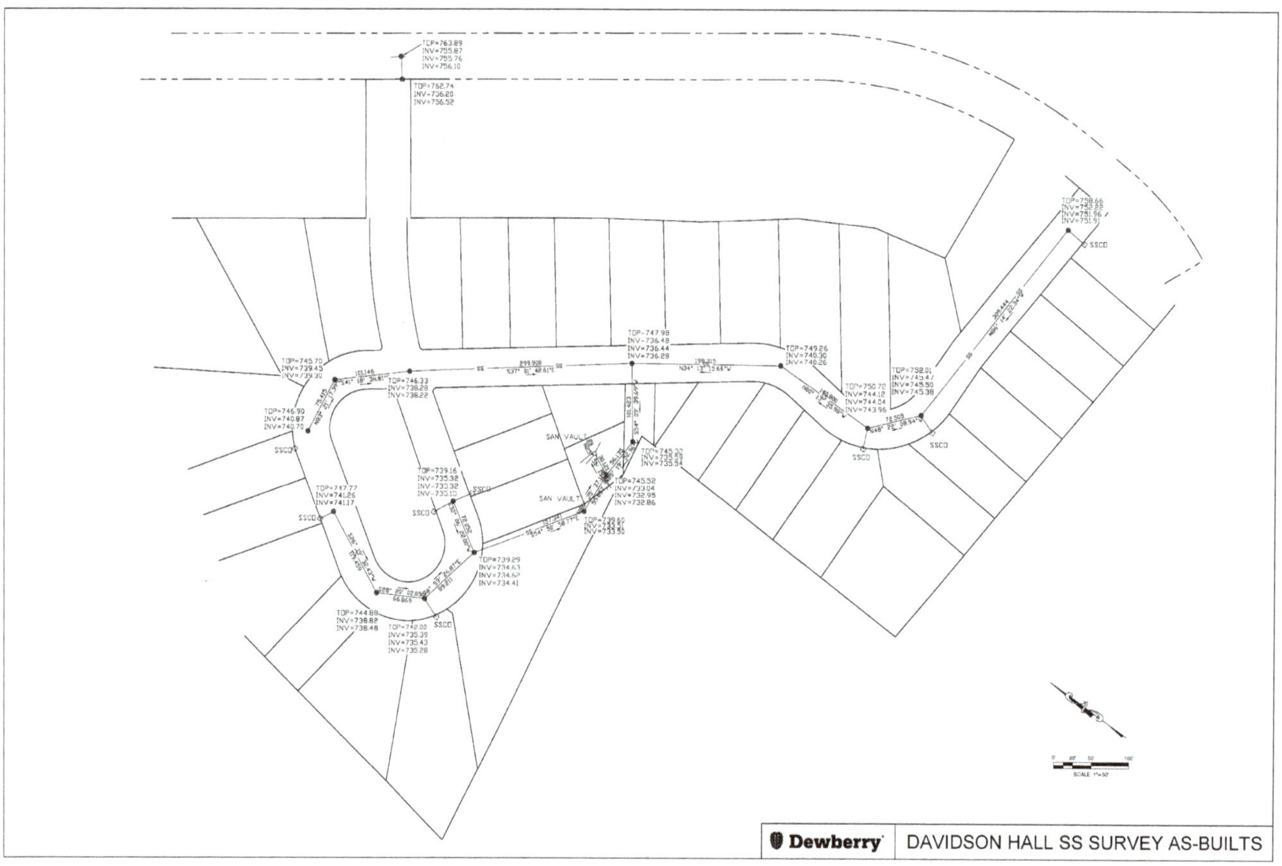

FIGURE 8.3 As-built survey for sanitary sewer.

5. **Roads—**Location of road, signage and traffic control devices in addition to finished grades. This as-built survey guarantees the road is built in accordance with the approved plans and is in the correct location with respect to the existing and proposed right-of-way.
6. **Parking—**Location, dimensions, and number of various types of parking including loading spaces, accessible spaces, and standard parking as required by the local jurisdiction's zoning ordinance.

Some local jurisdictions or public developers may require record drawings instead of as-built drawings. In this case, the civil engineer will receive the as-built drawings or survey from the contractor and incorporate the revisions into the construction drawings. One way a civil engineer may compile the as-built revisions into the construction drawings is by crossing through any outdated information on the construction drawings and noting the as-built condition (see Fig. 8.4). The local jurisdiction may also require any revisions to be shown in a different color, such as red. For any discrepancies between the as-built condition and the original design, civil engineers should review the adequacy of the constructed condition. For example, if an as-built survey reveals the constructed storm sewer pipe slope is flatter than the designed slope, they should check the hydraulic calculations to confirm the pipe is adequately sized at the new slope. If not, then they should notify the developer. The contractor may be required to rebuild a portion of the storm sewer to provide adequate drainage. Similarly, civil engineers should check minimum velocity calculations for sanitary sewer pipes. Record drawings require either a civil engineer's seal or certification statement, and thereby, the civil engineer shares some level of responsibility in the accuracy of the as-built information. For this reason, it is important they

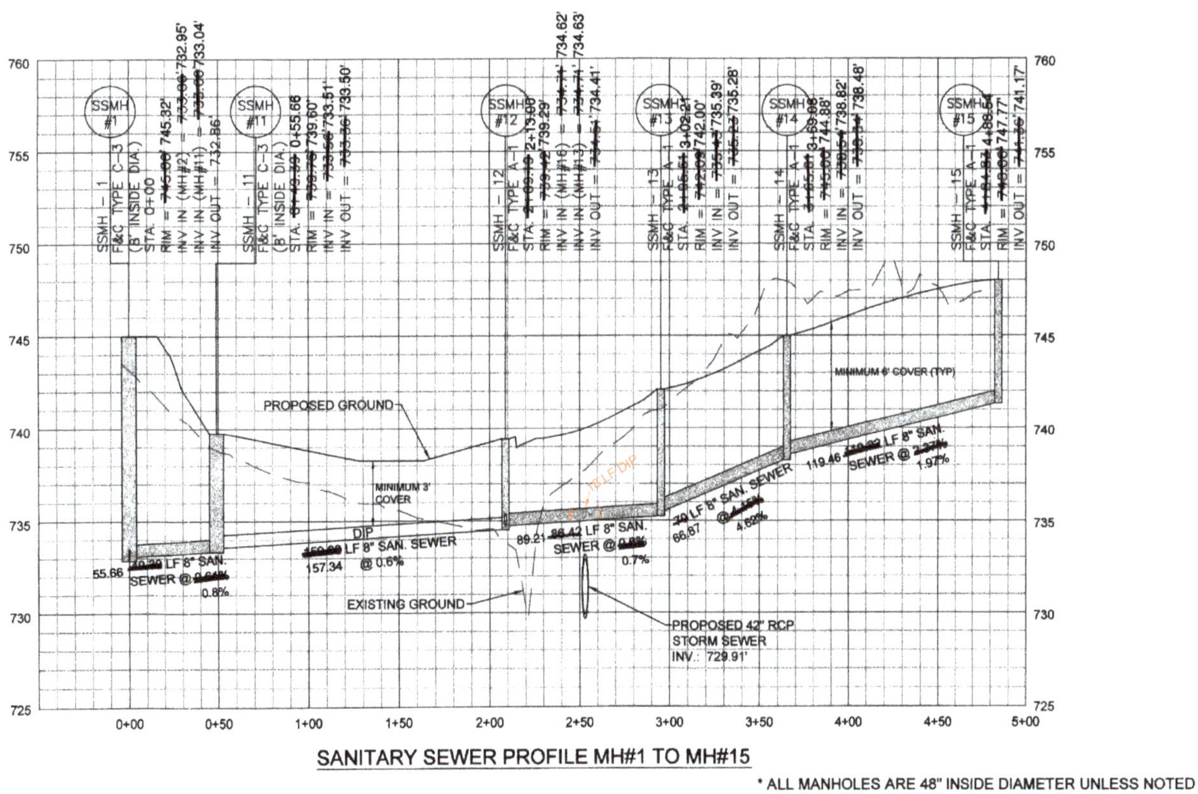

Figure 8.4 Record drawing of sanitary sewer profile.

note on the record drawings the source of the information and also include a disclaimer to limit their liability. An example of a disclaimer that appears on the cover sheet of record drawings is as follows:

> This drawing has been modified to reflect changes made during construction of the project. This is based on information from the contractor, the surveyor, and/or construction observation by the engineer's authorized representatives.

Even if a record drawing is not required, it is a standard practice for civil engineers to review the as-built drawings or survey. They should identify any of the as-built conditions that conflict with the construction drawings and thereby impact the long-term functionality of the site. As noted with the record drawings, this may require checking hydraulic or settling velocity calculations. Additionally, depending on the detail of information provided in the as-built survey, civil engineers may note any issues with the finished grade including accessible routes or inadequate slopes for site drainage.

8.4. ENGINEERING CERTIFICATION

Most local jurisdictions will require an engineer's certification for specified site work activities at the end of construction and prior to the issuance of the certificate of occupancy. Examples of certified site work activities include water distribution systems or stormwater management facilities. The building will typically have a final statement of special inspections from the structural engineer of record and/or the special inspection engineer of record. Engineering certifications typically include a statement indicating the construction adheres to the local jurisdiction's technical standards and specifications and is in accordance with the site plans approved by the local jurisdiction. Often the local jurisdiction will have a standard format for this certification that the civil engineer will update with project-specific information, sign, and seal.

In addition to the engineering certification required by the local jurisdiction, the developer may also require a certification stating the site work in its entirety was completed in accordance with the site plans and meets the local jurisdiction's zoning ordinance, standards, and specifications. Civil engineers serving in a typical construction administration role are not on the site at all times but instead visit the site at a regular frequency. Therefore, they should include appropriate disclaimers within the certification indicating the extent of construction observations. An example of such disclaimer is as follows:

> *Based on weekly construction observations and a final walk-through of the site, the construction was completed in general accordance with the site plans.*

An example of an engineering certification letter is included in Fig. 8.5.

DATE:	March 1, 2021
TO:	Washington County Engineering Department
FROM:	Suzie Jones, P.E.
RE:	Engineering Certification for the Main Street Recreation Center

This correspondence shall serve as certification that all engineering design and construction work on the above-mentioned project has been completed in accordance with all applicable technical standards and specifications and the construction plans approved by the Washington County Engineering Department.

Any variations from the originally approved construction plans are noted in the enclosed Record Drawings and do not substantially alter the original design.

Suzie Jones, P.E.

FIGURE 8.5 Engineering certification letter example.

8.5. CERTIFICATE OF OCCUPANCY

A local jurisdiction will issue a certificate of occupancy (C.O.) following final inspections confirming the building addresses all building codes, is in conformance with the approved plans, and is safe for inhabitance (see Fig. 8.6). Obtaining the certificate of occupancy is important to the developer as lenders typically require the C.O. prior to closing a loan. Often completion of site work activities is required to issue the C.O. These site improvements

Certificate of Occupancy
Washington County, California
Planning and Building Department
C.O. No.:

☐ Temporary ☐ Permanent

In accordance with §1234.56 of the Washington County Municipal Code, this document certifies the subject building or portion thereof was inspected and is deemed to be in compliance with the referenced code section and Washington County zoning, building, and use standards.

		Exceptions / Variances / Limitations:
Issuance Date:		
Effective Date:		
Expiration Date:		
Building Permit No.:		
Property Address:		
Owner Name:		
Owner Address:		

Construction Type:		Total Building Height (ft.):	
Building Type (New/Exist.):		Total No. of Dwelling Units:	
No. of Stories:			

	Use Classification	Area (s.f.)	Occupancy Group
Primary Structure:			
Accessory Structure 1:			
Accessory Structure 2:			
Accessory Structure 3:			
Accessory Structure 4:			

Internal Sprinkler System

System Required (Yes/No)?		Storage Height:	
Sprinkler System Type:		Aisle Width:	
Sprinkler System Location:		Sprinkler System Demand @ Riser Base:	
Hazard Classification:		Standpipe System Demand @ Base of Riser:	

Building Inspector	Building Official

FIGURE 8.6 Certificate of occupancy form.

often include setting the water meter, installing required landscaping, and grading the site in accordance with the local jurisdictional standards. Some jurisdictions will issue a temporary C.O. if all building components are complete and in accordance with the building code, while site work activities are in progress. A temporary C.O. will allow a tenant to complete interior finishes such as decorating or stocking shelves but will not permit full opening of the building for retail or residential use. A reinspection after all required site activities are completed is required to issue the full certificate of occupancy.

8.6. OPERATIONS AND MAINTENANCE

Record drawings do not represent the conclusion to land development consulting services; the long-term success of land development projects is dependent upon proper operations and maintenance (O&M). The final improvements will be managed by any number of entities including a professional management group, Homeowners' Association (HOA), or an individual homeowner. Regardless of the managing party, the civil engineer should clearly define and communicate the O&M responsibilities and procedures to achieve the design intent of the constructed facility.

Proper O&M procedures involve three distinct steps:

1. **Identify O&M responsibility**—often tied to ownership and/or location of the facility. It is important to *understand who is responsible* for carrying out long-term maintenance of each building and site component.
2. **Provide O&M procedures**—maintenance procedures vary depending on the specific facility or infrastructure system. At a minimum, *clarify what needs to be done* through directions for maintenance activities and a recommended schedule for completion. Disseminate this information to the responsible parties.
3. **Develop management tools to facilitate proper O&M**—accurate, informative mapping is vital to successful O&M. Mapping outlines responsible individuals and how to track or record O&M processes. The civil engineer likely has all the relevant information spatially referenced such that it can be *packaged in a manner that is useful to the designated maintenance personnel*. Whether it is tying the map to an online database or providing hard copies in the form of a comprehensive map book or atlas, the information must be readily available and in accordance with the management system employed by the developer.

Many jurisdictions require certain components of the O&M program to be specified on the plans and plats including specifying if the facility is public or private. Typically, a recorded easement will indicate public or private ownership. However, as easements are not visible in the field, signage, or other physical indicators are required in order to assess maintenance responsibility unless refined maps are available (see Figs. 8.7 and 8.8). Further, "privately maintained" can be a vague description particularly when multiple utility providers, HOAs, or other management entities are involved. For instance, a private utility pole may belong to the electric company, the cable company, the telephone company, or a HOA. Without a field indicator or accurate mapping, it is an iterative process to determine responsibility.

Occasionally, jurisdictions require maintenance instructions and criteria for certain facilities on plans or plats including stormwater management, landscaping, and onsite wastewater treatment facilities. In addition to these instructions on the construction documents, the jurisdiction may also require the developer to sign a maintenance agreement for any privately maintained facilities. A maintenance agreement allows jurisdictions with limited staff to enforce maintenance requirements upon private owners. Additionally, maintenance of these facilities is important as a part of low-impact development (LID) or sustainable development, which continue to become mainstream approaches to large and small-scale development. In fact, many of the sustainable certification programs—US Green Building Council's (USGBC) Leadership in Energy and Environmental Design (LEED) systems, National Association of Home Builders' (NAHB) Green Guide for Homes, and ENVISION—recognize and award credits/points for O&M training and education, monitoring, and post-construction facility optimization. By providing clear operating instructions, conducting maintenance staff training and occupant education, and disseminating accurate facility information (inherently collected during the process), the civil engineer can continue to assist developers in meeting their green building goals.

Facility management tools such as interactive maps are increasingly common given the advances in computer-aided design and drafting (CADD), which include smart drawings, building information modeling (BIM), and geographic information system (GIS) technology. The ability to spatially link referenced data to interactive maps,

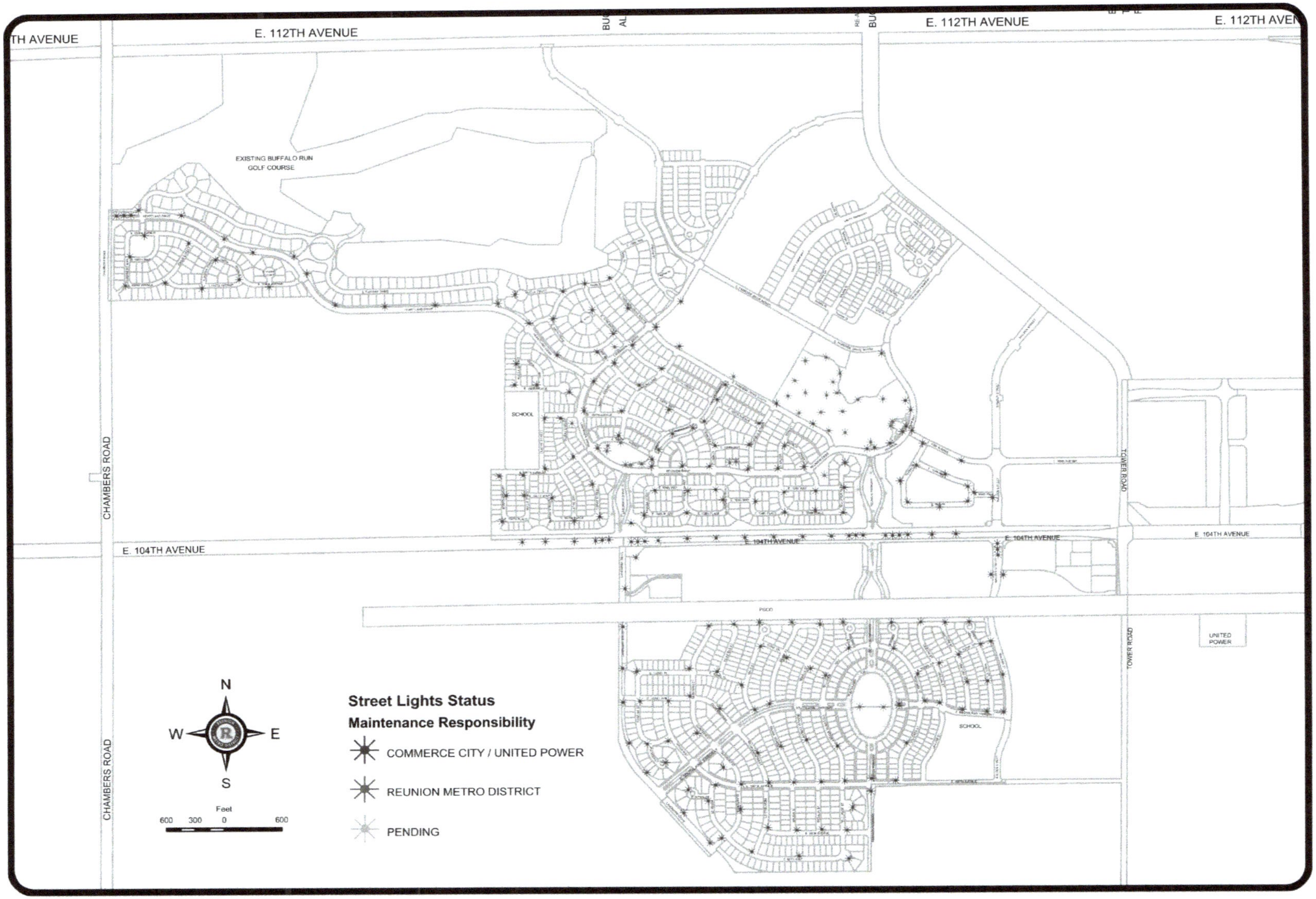

Reunion Metropolitan District

Street Lights

GIS Mapping Services provided by
GSI GEOMATIC SCIENCE, INC.
8000 S. Lincoln St. Suite 208
Littleton, CO 80122
303.703.4676

FIGURE 8.7 Reunion metropolitan district maintenance responsibilities.

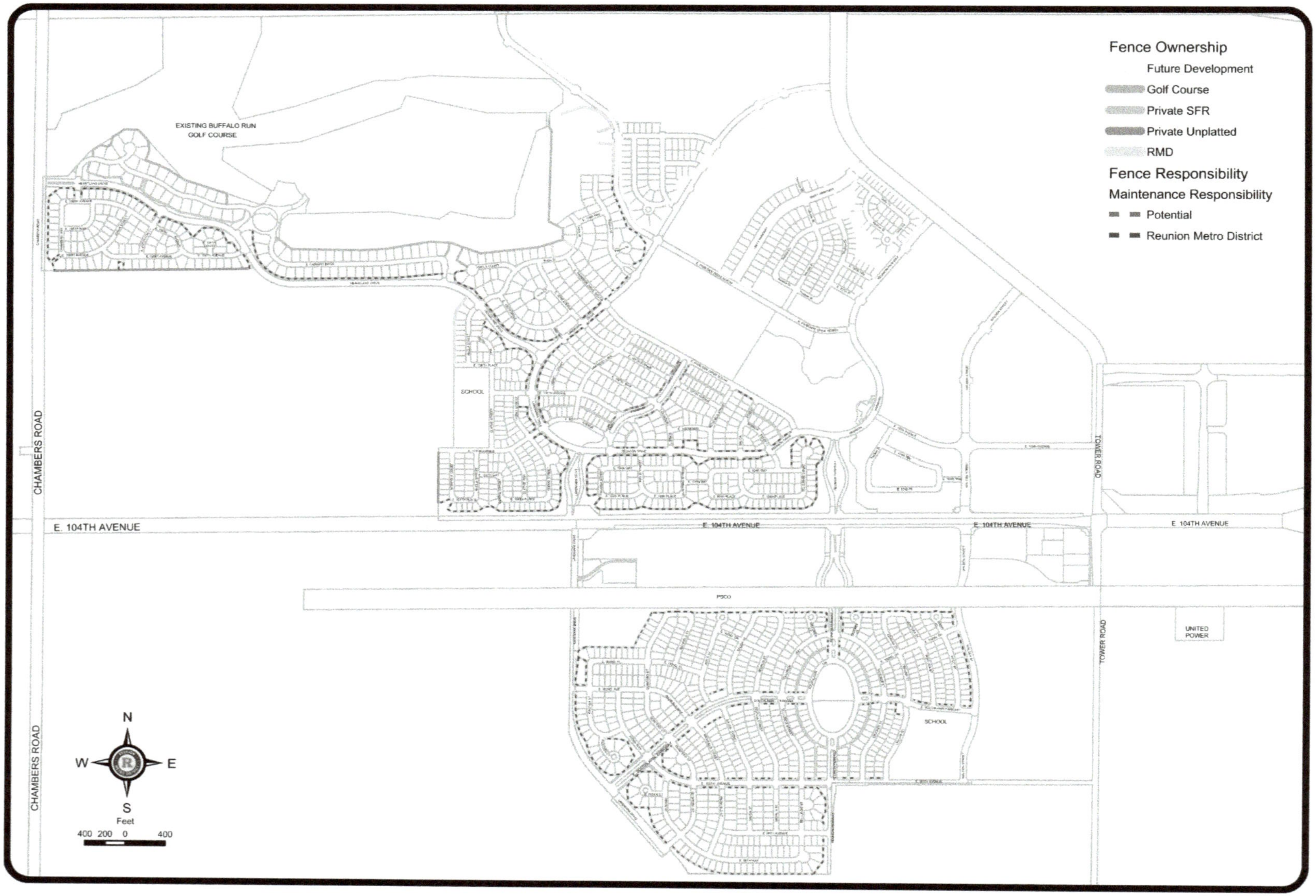

Reunion Metropolitan District

Fence Ownership & Maintenance Responsibility

GIS Mapping Services provided by GSI GEOMATIC SCIENCE, INC. 8000 S. Lincoln St. Suite 208 Littleton, CO 80122 303.703.4676

FIGURE 8.8 Reunion metropolitan district fence ownership and responsibility.

customized databases, online web resources, and query systems is the beginning of life cycle project management. This final deliverable may take many forms (digital or hard copy, manual or map book) and is created during the course of design and construction as information is developed, gathered, and assembled. This final packaging step is a means of consolidating all the design and surveying efforts into a single useful tool—an accurate record of the facility, its systems, and its O&M requirements. This can then be used as a way to track or log activity related to the successful function of the building and infrastructure systems.

CASE STUDY: ENGINEER'S COMPLETION NOTE

ISSUE:

Effect of engineer's recommendation of final payment, and owner's making of final payment, on owner's subsequent construction defect claim against contractor.

SUMMARY:

For a sanitary sewer project, EJCDC Standard General Conditions appear to have been used. The engineer recommended final payment, and the city made such payment. The city subsequently brought a breach of contract and breach of warranty claim against contractor, based on subsequently discovered defects. The contractor contended that by making final payment the city had waived the right to make a subsequent claim regarding the quality of the work.

The appellate court rejected the notion that engineer's recommendation was a binding determination that the work complied in all respects with the contract documents—it merely indicated that "to the best of Engineer's knowledge" the work had been satisfactorily completed such that payment could be made. The court also rejected the waiver argument, pointing out the clearly stated exception in the waiver clause (exception for claims for defective work, failure to comply with contract documents, or breaches of contractor's continuing obligations).

As a side note, the court expressed uncertainty as to whether a post-completion warranty claim must first be presented to the engineer.

COMMENT:

The court here had no trouble identifying and applying the substantial exceptions to the waiver set out in C-700 Para. 14.09.A.1, or in recognizing the related application of Para. 17.04, Survival of Obligations. As a matter of policy, EJCDC to date has viewed progress and final payments as merely that—payments—and not as validations of the work, acceptance of nonconforming work, or waivers of latent defects. If EJCDC considers a different approach such that payment does constitute a broad waiver, then owners are going to be reluctant to make final payment and release retainage. The format of 14.09 is such that the exceptions substantially outweigh the rule. The Construction Subcommittee is currently considering a restructure/rewording that would present the current exceptions as the rule. The issue of engineer's role in post-completion disputes is also under review.

Case study based on City of Kimberling v. Leo Journagan Construction Company, Inc. Missouri Court of Appeals (2011).

The information provided does not constitute legal advice. Consultation with an attorney is strongly advised.

8.7. CONCLUSION

Documentation of site improvements is an important aspect of the project closeout process relative to permitting and O&M. The developer is reliant upon the civil engineer to observe and note final construction deficiencies. This allows corrections to be made as needed before the contractor demobilizes and the developer releases the contract retainage. Verifying as-built versus designed conditions protects the developer and upholds public safety, as stormwater, sewer, and accessible areas are checked for adequate capacities and slopes. The documentation process is also an essential component of permitting and project sustainability. Certifying that site improvements are initially constructed properly can result in reduced O&M costs. O&M is an inevitable aspect of all constructed improvements, and it is a process made more efficient through effective documentation at project closeout.

CHAPTER 9

CONSTRUCTION STAKEOUT SURVEY

9.1. INTRODUCTION

Surveyors execute the mathematical calculations associated with the construction of the site and the field surveys that accurately reflect these calculations. Once the site plans are finalized and construction contracts are awarded, they then begin the construction layout process. The construction survey team is not always the same team that performed the boundary and topographic survey, and often the construction surveyor is a subcontractor of the general contractor, not the developer. In construction stakeouts, surveyors often provide one additional measure of quality control by evaluating the design intent of the plan documents and may catch design errors prior to performing a construction stakeout, preventing costly construction errors.

Three frequently used terms in construction surveying are *location, stakeout,* and *as-built.*

Typically, the construction stakeout sequence begins with clearing and rough grading stakeouts, followed by stakeouts for the utilities (sanitary sewer, storm drain, and waterlines) as well as the curb and gutter, building footprints, stormwater management facilities, pavement, and property corners.

9.2. CONSTRUCTION STAKEOUT FUNDAMENTALS

9.2.1. Stakeout Markers

For construction purposes, the stakeout marker of choice usually consists of a hub, a tack (or nail), and a guard stake. A hub is a 2 inch by 2 inch wood peg driven flush with the ground and marked on top with a tack. Surveyors note the horizontal and

Location refers to the type of survey where a specific existing feature is measured both horizontally and vertically from known control points. The data is then computed, plotted, and incorporated into the design.

Stakeout is the field process in which the surveyor provides in-ground reference markers for the contractor, which physically delineate the placement of utilities, streets, and buildings. Construction stakeout, or layout, is necessary in every phase of construction, from the initial clearing of the site through the setting of final property corners.

As-built (or record) drawings refer to a post-construction survey of a site, usually performed after the utilities are installed or the buildings are erected. As-built surveys, their importance and their specific requirements are discussed in Chapter 8.

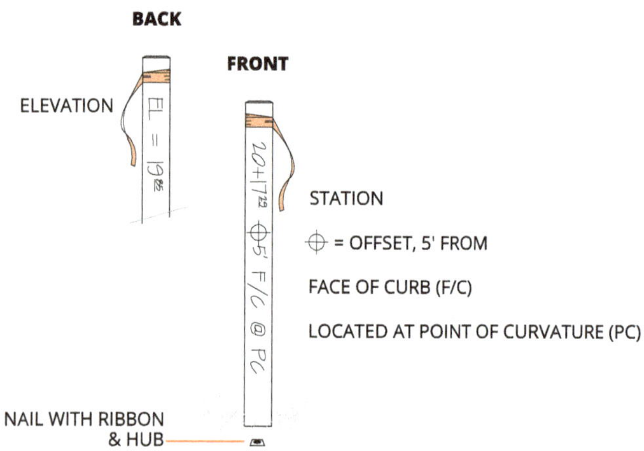

Figure 9.1 Hub and tack with guard stake.

vertical location of the tack in their field notes. A guard stake, made of plastic or wood, is a slat with one pointed end driven into the ground near the hub (see Fig. 9.1). It bears written information expressed in symbols, numbers, and abbreviations. The guard stake is placed approximately six inches from the hub to avoid disruption, such that the side with the offset distance faces the item being staked. Other types of markers used in stakeout include steel bars, spikes, nails with flagging or washers, chiseled marks in concrete, and drill holes.

Stakeout markers and accompanying information are used to indicate points of construction to the contractor but may also be used as reference control points for subsequent surveys. The hubs and guard stakes are often buried, dug up or otherwise disturbed on an active construction site. Cooperation between the surveyor and contractor is necessary to schedule the construction stakeout for a specific phase and to place the stakes in areas where they are not prematurely disturbed by other construction activities.

9.2.2. Offsets

Stakeout markers are typically set along lines parallel or concentric to actual centerlines, such as a road centerline, to preserve vital reference points outside of the actual area of work. The distance from the actual centerline to the stakeout line is the offset distance (see Fig. 9.2). The offset's distance is controlled by depth of excavation, width of the excavation machinery, the location of construction materials, traffic, and sometimes by governing authorities. Additionally, a contractor may request a specific offset distance in order to ensure construction activities will not disturb the offset.

Offset stakes are typically labeled in the following order:

- Distance of offset—labeled as left or right of the centerline based on the direction of increased stationing
- Identification of object being offset
- The station
- Vertical information—located on the back of the guard stake

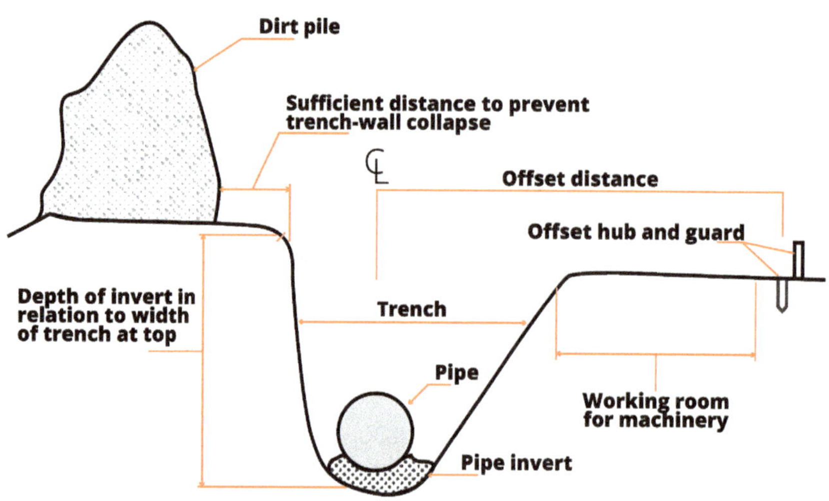

Figure 9.2 Open trench, pipe invert, and offset line.

AHD	Ahead	⊼	Instrument,	RF	Right Front
∡	Angle		Instrument Operator	RR	Right Rear
∡PT	Angle Point	INTX	Intersection	Ø	Rod/Chain Person
⌧	Area in Square Feet	INV	Invert	SAN	Sanitary Sewer
@	At	LF	Left Front	SAN LAT	Sanitary Lateral
BK	Back	LR	Left Rear	S/O	Stakeout
BM	Benchmark	⋈	Notekeeper	STA	Station
⊠	Benchmark	#	Number	STM STR	Storm Structure
℄	Centerline	O/S	Offset	ST	Street
COR	Corner	⊗	Offset	⊡	Square Point
V̲	Cut	-¦-	Planetable Setup	TAN	Tangent
Δ	Delta	PC	Party Chief	TBM	Temporary Benchmark
EP	Edge of Pavement	PC	Point of Curve	TB	Toe of Bank/Top of Bank
ES	Edge of Shoulder	PI	Point of Intersection	TS	Toe of Slope/Top of Slope
FC	Face of Curb	PRC	Point of Reverse Curve	TC	Top of Curb
⊼	Fill	PT	Point of Tangent	△	Traverse
⚹	Grade	POC	Point on Curve	⚠	Traverse Line
GB	Grade Break	POL	Point on Line	TP	Turning Point
Δ/2	Half Delta	PCC	Point of Compound Curve	VERT ∢	Vertical Angle
HI	Height of Instrument	PCTC	Point of Curve, Top of Curb	W/L	Waterline
Hor	Horizontal	PL	Property Line	WS	Water Service
HSE	House	RAD	Radius		

FIGURE 9.3 Surveying and mapping symbols and abbreviations.

Symbols and abbreviations are typically used as shown in Fig. 9.3.

9.2.3. Stationing

Stationing is a method of designating distances from known points on the site, usually measured in the multiples of 100-feet. The beginning point is typically designated as station 0, written as 0+00. A point 100-feet away would be station 1, written as 1+00. However, to avoid the use of negative stations (i.e. –1+50), the beginning station is often designated as an arbitrary full station other than 0+00 such as 10+00. Full stations are station at even multiples of 100-feet. Pluses are intermediate points between full stations. For example, a point 34.23-feet from station 1 is station 1+34.23 (pronounced as station "one plus thirty four point two three"). Stationing is the common method used on centerlines of streets and utilities as well as along the baselines of streams. For sanitary sewer systems, storm drainage systems, and streams, a common practice is to orient the stationing system such that the stations increase in the upstream direction.

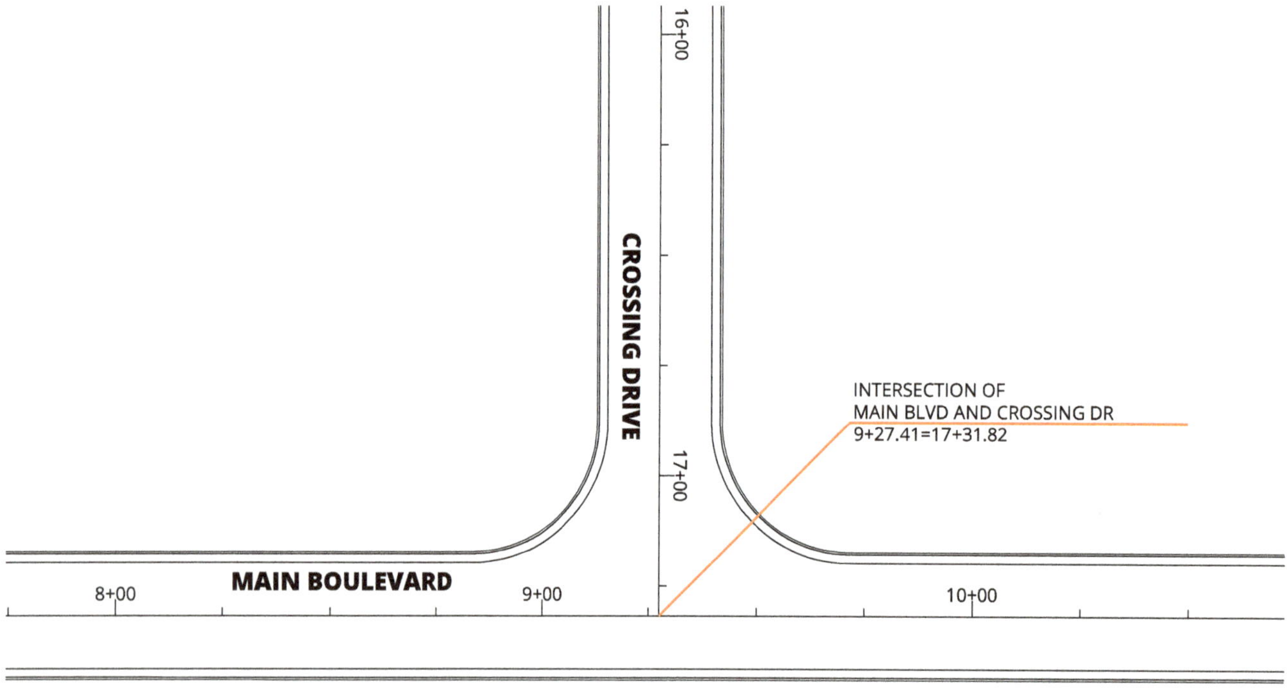

FIGURE 9.4 Example of street stationing.

An equality is used at points of intersections or to designate a change in the station numbering system. For example, a station equality is shown at the point where the two streets intersect, which identifies (i.e. equates) the station street to the station of the intersecting street. For example, Fig. 9.4 shows a station equality where two streets intersect, equating the Main Blvd. centerline station of 9+03.2 to the Crossing Drive centerline station of 10+00.

9.3. CUTSHEETS

A cutsheet organizes and presents stakeout calculations. Surveyors use cutsheets to record what is done in the field. Contractors use it to establish building and trench lines and grades while municipalities may require it as part of the permit and inspections process. A typical cutsheet form is shown in Fig. 9.5. Specific cutsheets can be tailored for a stakeout.

After establishing the stakeout and elevations for the hubs, the surveyor uses the field notes and construction drawings to fill out the cutsheet. The column labeled "STAKE EL." represents the offset hub elevation as determined in the field whereas "GRADE EL." is the proposed elevation of the pipe invert, manhole rim, storm grate, building floor elevation, top of curb or other desired point of interest. The grade elevation is taken from the plan or profile of the construction drawings, and the difference between the stake and grade elevations is the amount of cut or fill.

The surveyor computes the proposed grades for gravity utilities such as sanitary sewer and storm drain using the invert elevations and the slope of the pipe provided in the site plans. For the inverts of pressurized pipes such as waterlines or sanitary sewer force mains, the surveyor uses the elevations shown on the profile in the site plans or subtracts the required minimum cover over the pipe from the finished grade elevation.

The surveyor also includes top of curb or edge of pavement elevations for road and parking areas in the cutsheet calculations. Top of curb elevations for roads are easily determined from the longitudinal grades and the typical road cross section. However, as parking areas have variable and nonuniform longitudinal and cross slopes, the surveyor uses spot elevations from the grading plan or judgment based on the provided contours to determine top of curb elevations. The surveyor may request additional spot elevations from the civil engineer if the plan is too vague.

County #: ______________________ Plan #: ______________

Project : ______________________ **Grade to:** ______________________

____

Prepared by : ______________ **Date:** ______

Staked by : ______________ **Date:** ______

Dewberry®

8401 Arlington Boulevard
Fairfax, VA 22031-4666
703/849-0229 Tel
703/849-0118 Fax

DESCRIPTION	OFFSET	STATION	STAKE EL.	GRADE EL.	CUT/FILL	REMARKS

FIGURE 9.5 Typical cutsheet.

The cutsheet is one of the last points in the design process where the development team can catch errors and conflicts before construction begins and therefore, before revisions become more costly to correct. The person computing the cutsheet should watch for conflicts where pipe and other utility crossings occur, confirm minimum depths are met, and identify other possible engineering, field, and construction conflicts. Typically, it should take three days or less from the completion of the field stakeout to prepare the cutsheet and return it to the contractor to begin construction.

This chapter will further outline descriptions and examples of cutsheets for particular stakeouts in the following sections.

9.4. LEVELING METHODS

The offset hub elevations establish a vertical reference for the contractor's use. Most surveyors use trigonometric and global positioning system (GPS) leveling to obtain construction stakeout elevations, but they can also use differential leveling and three-wire leveling.

9.4.1. Trigonometric Leveling

Trigonometric leveling uses a total station instrument (TSI), prism pole, and data collector. It applies the properties of a right triangle to determine differences in elevation. As shown in Fig. 9.6, the vertical angle between the TSI's horizon line and an intercept of the rod or prism is measured. The slope distance is also measured. With this information, the surveyor can calculate the vertical distance between the rod intercept and horizon line, which can be positive or negative as shown in Fig. 9.6. The ground elevation is calculated as the sum of the height of instrument (HI) and the calculated vertical distance, subtracting the rod intercept height measured on the prism pole.

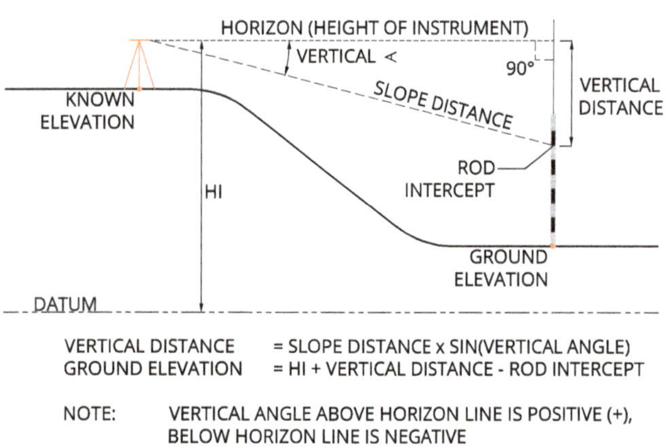

Figure 9.6 Trigonometric leveling.

9.4.2. GPS Leveling

GPS equipment can be used to directly determine the elevation of any point such as control points. Using a GPS requires multiple observation periods of 30 minutes or longer. A real-time kinematic (RTK) system consists of two GPS units: a base station and a rover. The base station is placed on a prominent, high quality control point where it will remain while the rover unit collects data.

Typically, each site has a designated control point used by the base station. For continuity throughout the project, the base station should use only one control point. The rover is free to move about the site to perform the necessary location or stakeout survey. The two units constantly receive GPS signals and exchange information with each other via radio signals or cell phone transmissions. The rover, equipped with the data collector, can then combine both GPS signals to produce high-quality data in both the horizontal and vertical direction. The longer the observation period for the rover, the greater the accuracy of the horizontal and vertical location.

The virtual reference systems (VRS) is the next extension of the RTK concept. VRS is a series of GPS reference stations, located throughout a relatively large geographic region (such as a county, state, or group of states), and linked together by a software solution via the internet. The rover also connects to the system via the internet using wireless technology and provides real-time data and corrections from the network of GPS reference stations. Rather than the single or site anchored base station used in traditional RTK surveying, this system establishes a virtual reference station on the job site.

9.4.3. Differential Leveling

Differential leveling consists of running a level circuit between a minimum of two established control points and obtaining rod readings on the hubs as the circuit progresses. The level circuit itself is

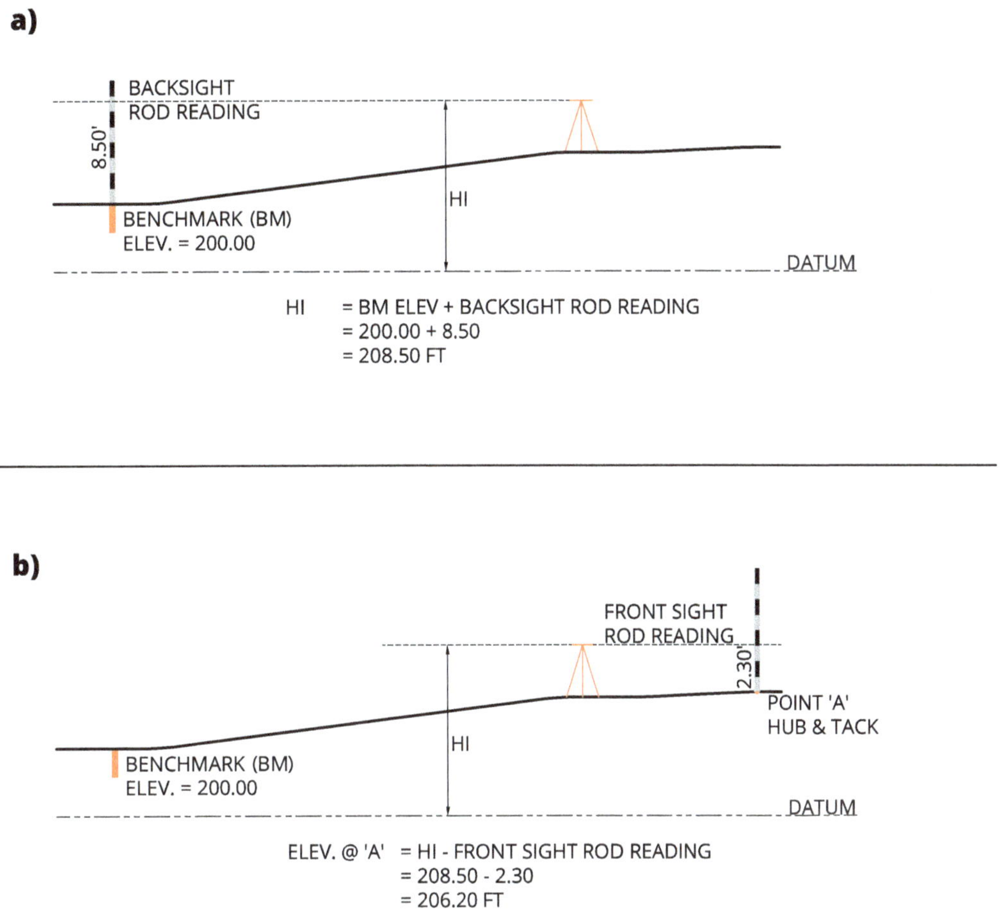

Figure 9.7 Establishing height of instrument (a) and determining elevation (b).

a series of adding and subtracting rod readings to obtain differences in elevation. Once the HI is established, any subsequent rod shots are subtracted from the HI to determine the elevation of that point (see Fig. 9.7).

In construction stakeout surveys, the observations of several offset hubs may be made from one instrument set-up. However, these elevations are technically unproven since they are side-shots and not a contiguous part of the run. Therefore, key points of the stakeout are turned upon during the level circuit. A turn consists of reading the rod on a stable point, changing the height (and location) of the instrument as needed, and reading the rod again (see Fig. 9.8). As part of a continuous level run, the turn is an integral arithmetical component and is proven by the final check-in at another known benchmark.

Successful turns should be balanced, roughly using the same distance for backsight and foresight, to eliminate any adjustment error in the instrument. When reading high on the rod, the rod person can gently rock the rod back and forth in the direction of the instrument allowing the instrument operator to read the smallest gradation on the rod, thereby eliminating the error created by an out-of-plumb rod.

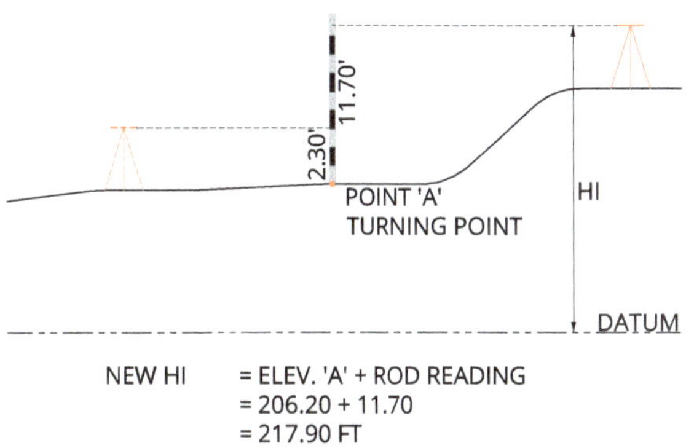

Figure 9.8 Establishing HI at a turn.

When jobsite benchmarks are established, they must be set in practical locations unlikely to be disturbed by construction activities. They must also possess an accurate elevation, proven by inclusion as a turning point in a level circuit and producing an acceptable closure. Recommended benchmark locations include a nail or spike driven into a preserved tree, a corner of an existing storm structure, a flange bolt on a fire hydrant, or a re-bar or pipe driven flush into the ground.

9.4.4. Three-Wire Levels

When greater accuracy is necessary, a surveyor may use a three-wire level. Many leveling instruments are equipped with three horizontal cross hairs. The top and bottom cross hairs are often called stadia hairs or wires. The middle hair at the center of the scope represents the horizon line of the instrument. On most internal focusing instruments, the spacing of the stadia hairs creates a proportion of 1-foot equal to one hundredth of a foot intercepted on the rod. In three-wire levels, the surveyor reads the rod at each intercept: top stadia hair, middle cross hair, and bottom stadia hair. The readings are made to the thousandth of a foot by visually dividing a hundredth on the rod into ten parts. The surveyor adds the readings together and divides by three to obtain the mean value. The mean should equal the middle cross hair reading, or be within one thousandth of a foot difference, to be valid. Good results may be obtained if only the first section of the rod is used (due to possible movement in the rod section clamps), each shot is balanced, and the distance from level to rod does not exceed 80-feet. The surveyor keeps notes using the same procedure as differential leveling (see Fig. 9.9).

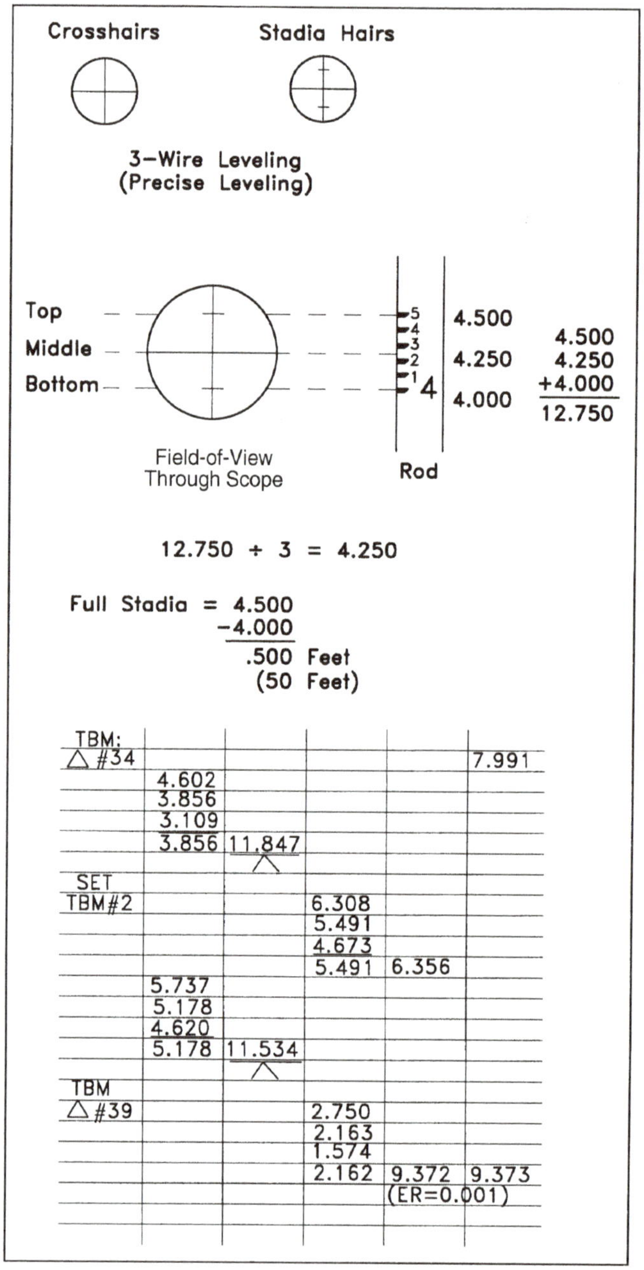

FIGURE 9.9 Three-wire level.

9.5. BASIC PROCEDURES

9.5.1. Field Procedures

Onsite, the survey party will do a reconnaissance of the area, looking for the traverse stations, property corners, and any other known horizontal control points. The most common method of distance measurement in construction stakeout uses the total station instrument. Past methods of distance measurement include pacing, chaining, and taping. These older methods are still used to locate property corners, traverse stations, and other points of interest during site reconnaissance.

After a reconnaissance is complete, the party chief must form a logical sequence for the survey, based upon the points recovered, stakeout desired, onsite obstacles, and the distances involved. While a timely survey is desirable, accuracy is a necessity. The party chief should have a firm grasp of trigonometry and coordinate geometry due to the number of computations solved in the field, such as inverses using coordinates or bearings and distances and intersections using bearings and/or distances. This knowledge is necessary to troubleshoot possible problems in the plans, create alternative ways to perform a survey, and to provide mathematical checks and corrections to work in progress. Data collectors are invaluable in aiding the party chief with these calculations.

The first step in a stakeout survey is to locate a minimum of three control stations to layout and check a survey. Ideally, the instrument occupies one control station while another control station is sighted. Then the surveyors measure a computed angle and distance to the beginning point of the stakeout. The instrument is moved ahead to the second control station point, and the first control station is backsighted. A new angle and distance is measured to a third control station. If this checks, then the accuracy of the beginning point is established. For a lengthy stakeout, the instrument is moved forward to each subsequently set point, backsighting the last, so that the whole series is proven to a known control station at the final check-in.

9.5.2. Errors in Stakeout Surveys

Errors in stakeout surveys often result from weather, equipment, terrain, and human fallibility. A survey party chief knows the tolerances of acceptable error. For example, stakes set for controlling earthwork activities need only to be within 1-foot horizontally of the plan location, whereas stakes set for house construction must be accurate to the nearest hundredth of a foot. The surveyor will determine the amount of tolerable errors based on the stakeout's purpose, how will it be used, and if the points will be used for another survey. If the tolerance is exceeded, the source of error must be found and corrected.

After the stakeout is complete, the survey crew reviews the stakeout in its entirety and compares it to the construction drawings. This overall review can reveal obvious errors that can be addressed prior to the start of construction.

9.6. CLEARING STAKEOUT

On a wooded site, the first stakeout requested typically establishes the clearing limits as depicted on the site plan. Limits of disturbance and clearing limits may be used interchangeably. The clearing limits and limits of construction however may be different as the construction limits may include areas for material laydown, construction trailers, or onsite parking outside the limits of disturbance. Establishing the clearing limits is critical as work performed beyond the permitted construction limits may have legal implications. Construction limits are often set at property lines and outside of environmentally sensitive areas. Working beyond the limits can cause irreversible damage to adjacent owners and environmental areas.

Surveyors typically perform clearing stakeouts using a total station instrument and prism pole with a data collector. On fully wooded sites, GPS is undesirable because the tree canopy blocks the signal and significantly degrades the survey accuracy. The only opportunity to use GPS for a clearing stakeout would be a partially wooded site during winter months. The limited canopy cover during these months allows enough of the GPS signal through to obtain the desired accuracies.

Coordinates are computed for points along the clearing line, and the data is uploaded into the data collector. In the field, the data collector provides the angles and distances from the occupied station to any point desired. The points that are set can be located and stored in the data collector and plotted in the office later, serving as a check of the work.

In the field, the survey crew recovers control points if they appear threatened by the clearing operation. The rod person flags each clearing limit point. The person at the instrument must confirm the points are spaced at intervals visible through the woods and protected against the loss of accuracy caused by spacing the tie points too far apart in a curve. After the key points are set, they are connected by running flagging between them. It is tied high and secured well so it will remain undisturbed. In areas where vandalism may be a problem, the contractor may request that the trees be marked with paint as well. Flagged trees along a limit line are not disturbed.

If right-of-way or easements are requested for a clearing stakeout, the clearing limits are measured at a distance to the left or right of the staked centerline points.

9.7. GRADING

9.7.1. Street Grading

The road centerline is generally used to control the horizontal and vertical limits of roadway construction. For subdivision streets, the stakeout consists of setting hubs at the centerline's geometry points such as points of curvature (PC) and points of tangency (PT) as well as at 50-foot intervals. As a check, ties measuring angles and distances are made to control points from selected PC and PT locations. The ties are made when control points can be sighted and measured from the selected points on the centerline. However, as the stakeout progresses, ties may be available infrequently. Typically, early control points are near the project's boundaries, and therefore, street stakeout ties are made near the boundaries.

The outside limits of the grading for the street are established by measuring, radially or perpendicular, from the centerline hubs the necessary distance. In most cases, the horizontal accuracy required for setting the outside limits for the rough grading is approximately 6 inches. The stakeout for the street is set on the actual centerline rather than offset. Although this might obstruct the equipment operators as they grade around these points, the close proximity of the points allows the contractor to set the street grades accurately even in the rough grading phase. Evidence of the saved points within the grading area are apparent on construction sites with the numerous cylindrical mounds with guard stakes and surveyor's flagging on top. When the street grading is complete, these mounds are easily removed.

Elevations are determined for all the hubs, starting at the point where the street construction begins. On most projects, street construction begins where the intersection of the proposed street meets an existing street near the boundary of the site. From here, the construction proceeds toward the interior of the site. In addition to providing a beginning point where the elevation of the proposed street matches the elevation of the existing street, this practice provides a logical order for surveyors to check the work against the plans and supplies the contractor with a cutsheet starting at the most accessible beginning point. If the cutsheet provides cut and fill depths to the finished grade of the street, the contractor is responsible for deriving the subgrade elevations.

9.7.2. Slope Staking

Engineering projects such as roads, rail beds, ditches, and canals often use slope stakes to define the limits of excavation and grading. The slope stake is offset from the intercept or "catch point," which is where the finished side slope meets the original ground surface. Slope stakes are especially helpful to the contractor in areas of cut where the catch point is less apparent than in areas of fill.

To begin the survey, the surveyor stakes the centerline of the road or ditch at regular intervals and determines an elevation at each station. To set a slope stake at each station, the surveyor uses a level, a leveling rod, and a measuring tape. The objective is to find the catch point left and right of each station using Equation 9.1 and potentially trial and error:

$$d = \frac{w}{2} + hs \tag{9.1}$$

where: d = the measured horizontal distance from centerline to the catch point
w = the plan width of the roadbed or ditch
h = the cut or fill at the centerline
s = the side slope ratio

These parameters are shown in Fig. 9.10. As shown, the inherent horizontal error in Equation 9.1 is due to the variation in depth, h, at the centerline and the actual depth at the catch points, h_l and h_r. For this reason, setting of slope stakes in the field becomes a trial and error process.

To begin the process, the surveyor positions the level and establishes the HI by taking a rod reading on the centerline station previously established. The surveyor extracts the final road grade elevations along the centerline and the typical cross section side slopes from the site plans to complete the calculations. The difference between the HI and the finished grade elevation is known as the "grade rod." The difference between the rod reading at a trial location and the HI is known as the "ground rod." The ground rod is variable and depends on the trial location whereas the grade rod is constant for a given centerline station. The cut or fill at any trial point in the section is figured by subtracting the

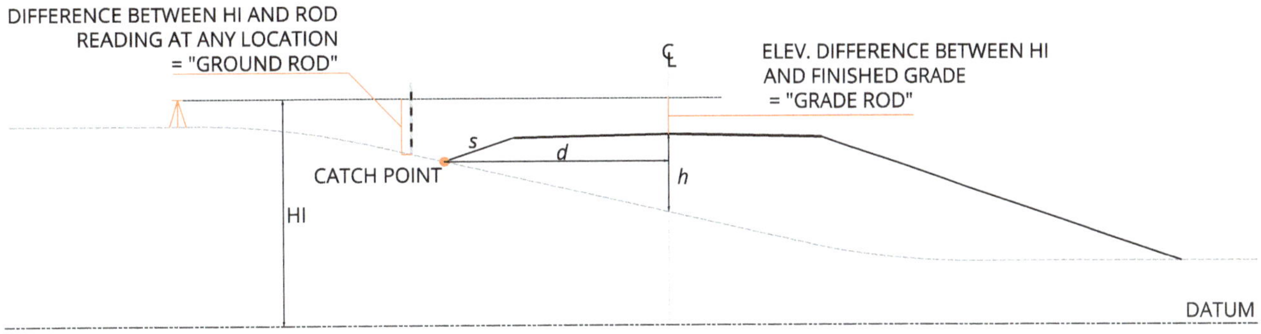

Figure 9.10 Slope staking.

ground rod from the grade rod. The equation can then be expressed as:

$$d = \frac{W}{2} + s\,(\text{grade rod} - \text{ground rod}) \tag{9.2}$$

The objective in the field is to find a ground rod such that the right side of Equation 9.2 is equal to the distance from the centerline at the location of the rod reading.

Example:

In this example, the road width is 24.0-feet, the side slope ratio is 2:1, the finished grade elevation at the centerline is 29.0-feet, and the HI is at elevation of 42.0-feet.

1. The grade rod is 42.0 ft – 29.0 ft = 13.0 ft.
2. Using Equation 9.1, the first trial distance from the centerline is 38-feet.

$$d = \frac{24}{2} + 2(13) = 38\text{-feet} \tag{9.3}$$

At a trial distance of 38-feet, the rod reading is 4.2. Using Equation 9.2:

$$38\text{ ft} = \frac{24}{2} + 2(13 - 4.2) \tag{9.4}$$

As the equation is not balanced, the left side of the equation must decrease and the right side must increase. In this case, the rod person is directed downslope or closer to the centerline.

3. The second trial, at 29.6-feet from centerline yields a rod reading of 4.3, which when placed in the equation shows: 29.6 = 12 + 2 (13.0 – 4.3). The right side of the equation equates to 29.4-feet, which does not equal 29.6-feet.
4. Moving the rod to a location of 29.5-feet from centerline produces a ground rod of 4.3-feet. This final trial yields 29.5 = 12 + 2 (13.0 – 4.3), thus, 29.5 ≈ 29.4. This is within 0.1-feet of where the rod is being held, so the catch point is recorded at 29.5-feet from the centerline and the depth of cut is recorded as 8.7-feet (13.0 – 4.3). The catch point is then offset, away from centerline, and the stake is marked to express the information to the contractor. The front of the stake indicates the offset distance to the catch point, the cut or fill from the catch point to the finished centerline grade, and the station. The back of the stake indicates the offset distance to the centerline and the cut or fill from the offset elevation to the finished centerline grade (see Fig. 9.11).

Construction plans for many subdivision streets contain a contour grading plan or street cross sections. If the topographic maps used to create these are reasonably accurate, the surveyor can calculate the catch points directly from these drawings.

9.7.3. Overlot Grading Stakeout

There is often a considerable difference between existing grades and proposed grades throughout a site. Consequently, the contractor hauls dirt from one place on the site to the other. Overlot grading is where the contractor mass grades the site to bring the entire site closer to the finished grade. This includes moving dirt onsite as well as importing additional fill or exporting excess material offsite.

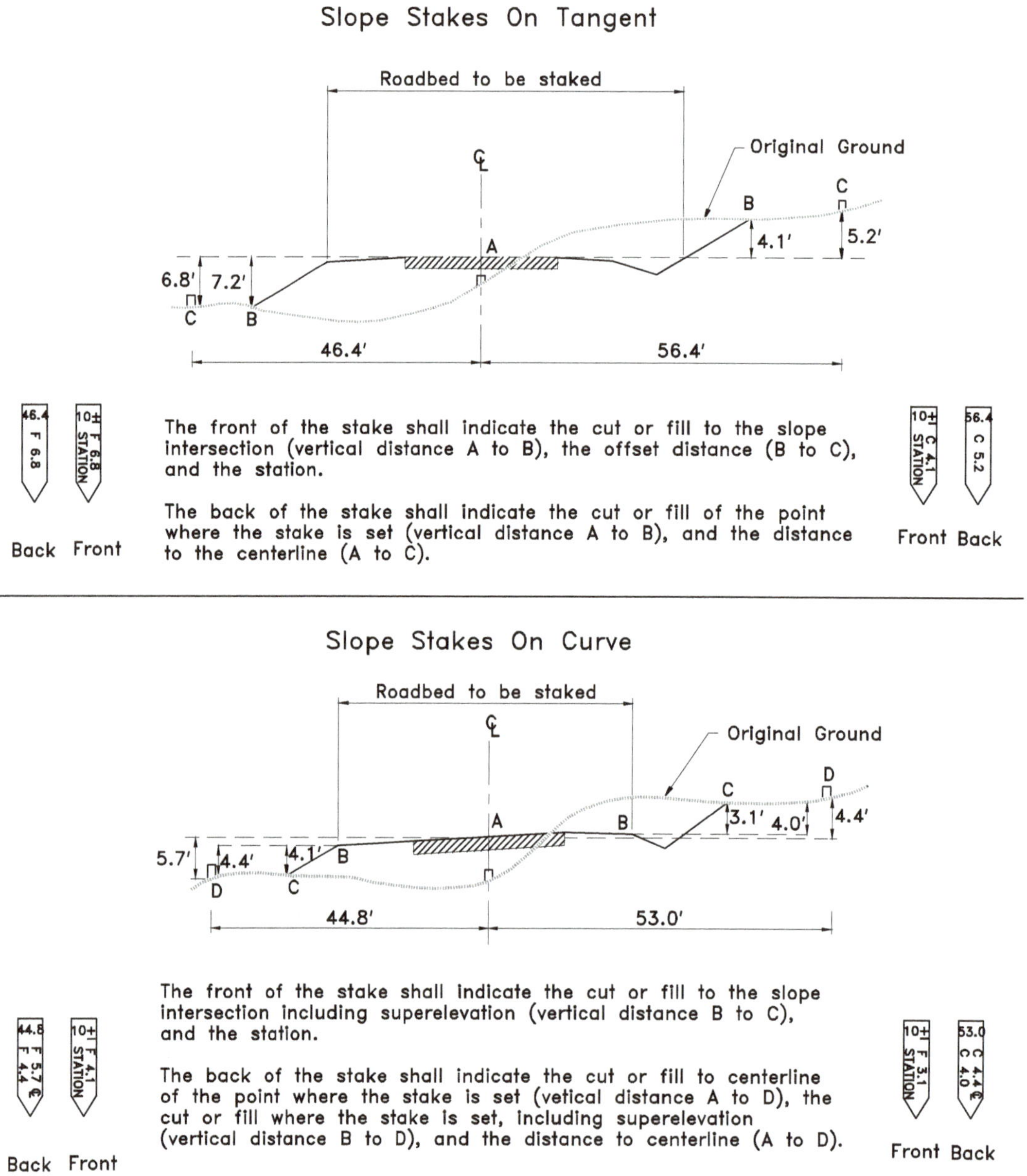

FIGURE 9.11 Setting and marking slope stakes.

For projects requiring a large amount of earthwork, an overlot grading stakeout may be requested. A surveyor uses a grid or contour stakeout to complete the overlot grading stakeout.

Grid Stakeouts. Surveyors use grid stakeouts when the earth base must be relatively accurate (i.e. ballfields). To ensure a reasonable degree of accuracy, they stake grid points at a close interval. Depending upon the size of the project, 25- to 100-foot grid intervals can be used, and cut or fill is determined for each grid point. The grid-point elevations are interpolated between contours, spot elevations, or building corner grades.

Contour Stakeouts. A surveyor uses contour stakeouts for ponds, berms, and slopes when less accuracy is required. A contour stakeout transfers selected contour lines from the plan to the field to guide the earthmoving operations. When selecting points, the surveyor considers their spacing and location in relation to the maneuverability of the grading equipment. The stakeout markers should be highly visible to the equipment operators, sometimes necessitating the use of 4-foot tall stakes guarding the hubs. Tall stakes also provide ample room for cut, fill, and grade markings, the proposed contour elevation, and the elevation of the hub.

Surveyors can use a total station instrument for the stakeout. After the hubs are set, they obtain elevations and issues a formal cutsheet with the cut and fill information.

9.7.4. Machine Control Systems

Advances in the use of Global Positioning Systems (GPS) have revolutionized the construction stakeout process. Specifically, the advent of Real-time Kinematic (RTK) GPS allows surveyors to perform construction stakeout surveys with unprecedented speed and accuracy. Contractors also utilize RTK units attached directly to their earth moving equipment such as graders or bulldozers. This is called machine control. The on-board RTK unit can inform the operator where more cut or fill is required to achieve the proposed ground elevations. This technology reduces or potentially eliminates the role of the surveyor for basic things like mass grading. However, a surveyor will still need to provide "as-built" information on the grades at critical points after the contractor completes machine control work to validate the work performed. Use of GPS systems on construction equipment can increase accuracy and efficiency of grading operations for both roadways and mass grading activities.

Machine control typically uses a digital terrain model (DTM) of the proposed grading to automate the grading activities. In a semi-automated system, the equipment operator may review the elevation information on a screen in the machine cab and manually adjusts the blade of the dozer or grader to match the plan. Alternatively, with an automated system, the operator maneuvers the equipment as indicated on the screen, and the blade is automatically adjusted by the equipment's hydraulic system to match the proposed elevations. The proposed grade DTM created by the civil engineer is different than the DTM used by the contractor for machine control. The civil engineer's DTM represents finished grade, such as top of curb or pavement grades. However, the contractor's DTM is used for mass grading and therefore built based on subgrade elevations.

9.8. CURB AND GUTTER STAKEOUT

Curb and gutter is staked from accurate control points and referenced to the street centerline. The main geometric points including PC and PT are usually reflected in the layout of the curb when the curb is parallel or concentric to the centerline. Other stakeout points include stations where street width transitions begin and end, intersecting streets, street entrances to commercial properties, high and low points, and stations corresponding to curb returns. The surveyor transfers the stakeout points from the centerline perpendicularly to the curb along tangents or radially along curves at the designated distance to face of curb plus the offset distance. Seven- and ten-foot offsets are commonly used as this is often the distance to the property line and convenient for use with a slip form paver used to construct the curb.

For a curb return at a street intersection, a perpendicular line from the centerline tangent or a radial line from the centerline curve also sets the radius of the return. Fifty-foot intervals are adequate along tangent sections, while 25-foot intervals are used through horizontal and vertical curves. Points along a vertical curve on the curb-offset line are set the same way as staking out a horizontal curve.

9.8.1. Curb Offsets through Street Transitions

Street transitions occur when the street changes width between intersections due to the addition of turn lanes. The transition occasionally occurs only on one side of the street or intersection. Transitions also occur with the addition of a lane to increase traffic capacity.

A smooth transition design can simply be two reverse curves. The beginning, midpoint, and end of transition points are set at their stations along the street centerline. Then each transition point is offset radially behind the face of curb at the transition widths noted in the plans. If the curves are equal in length, the distance to the midpoint, which is the point of reverse curve, will be half the distance between the beginning and end transition points.

9.8.2. Cutsheets for Curb and Gutter

Horizontally, curb offsets are always staked in relation to the face of curb. The top of curb is frequently used as the vertical reference point. The proposed street grades can be determined from the profile and the typical section. The surveyor compares this grade to the existing elevation from the stakeout notes and enters the corresponding cut and fill depths on the cutsheet.

Where the proposed curb connects to an existing curb, a minimum of three top of curb elevations along the existing curb at 25-foot intervals are obtained as the stakeout is run. This provides the cutsheet preparer with the information needed to confirm the newly constructed curb matches the existing curb. If the proposed grades do not match the existing grades, the cutsheet preparer should consult with the civil engineer. Otherwise, the surveyor can make slight adjustments to the elevations to provide a smooth vertical connection between the existing and proposed curb and gutter.

A frequent problem when preparing curb and gutter cutsheets for parking areas is misinterpreting the elevations listed on the site plan. Whereas the proposed top of curb elevations can be provided in the street plans and profiles, curb elevations may not be as readily available for areas with nonuniform geometry, such as parking areas. Rather than a typical cross section and profile as prepared for roadway grading, parking grading plans often use spot elevations in conjunction with grading contours. Drainage problems or uneven curb and gutter sections may result if spot elevations are too sparse or are not properly interpreted by the cutsheet preparer.

As an example, the typical curb and gutter section is not consistent for all site plans. It can vary based on the local jurisdiction requirements, type of curb and gutter selected, or based on developer preferences. Variations may include gutter pan width, slope, and curb height. These variations will all impact the elevation difference between top of curb and edge of pavement grades, and therefore need to be carefully reviewed during stakeout.

9.8.3. Curb and Gutter Stakeout in Parking Areas

Parking lot curb and gutter stakeouts are generally performed as previously described and shown in Fig. 9.12. However, additional items to consider include travel lane widths, parking bays, islands and medians, and accessible ramps as shown on the site plans.

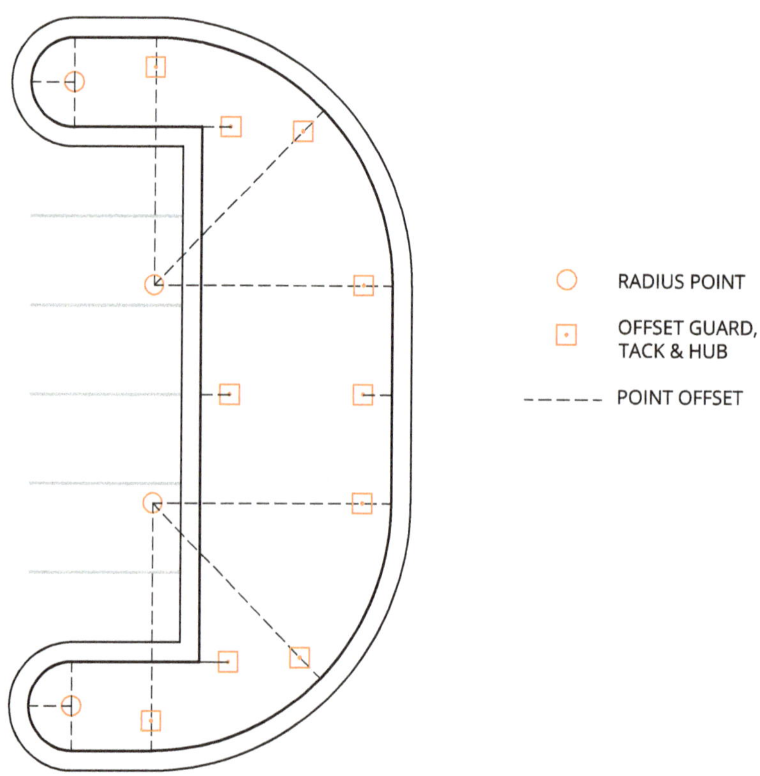

Figure 9.12 Parking lot curb island stakeout example.

Parking bays contain a certain number of parking spaces, usually noted on the site plan along with the width of each space. The number of parking spaces multiplied by the width of each space equals the distance between the two corners of the bay, and the surveyor should compare this distance to the stationing system. If there are any handicap parking spaces, the accessible aisle width is added to the total distance.

Parking lot islands channel traffic, provide landscaping areas, and accommodate central features such as community mailboxes, light poles, fire hydrants, and entrance gates. If an island incorporates utilities such as fire hydrants, storm drainage structures, sanitary sewer manholes, or light poles, the surveyor should check both the horizontal and vertical location of these features relative to the island.

Finally, the location of flush curb adjacent to accessible ramps, crosswalks, public trail access points, or loading zones may be included in the stakeout. A careful study of the site plan will reveal accessible routes and a record plat will aid in the siting of trails within easements.

9.9. BUILDING STAKEOUT

Two important considerations in establishing the building location for the stakeout survey are adherence to zoning setback regulations and allowable accuracy for construction. Although the site plans show the building conforming to setback requirements, the stakeout computations and the subsequent field procedures must ensure the field location matches the plan location. Additionally, the dimensions of the building and the corner angles must be set to within the tolerable limits required for construction purposes. Building stakeout can be as straightforward as setting offset stakes for the four building corners or as complex as setting column line locations for a high-rise structure which require a high level of accuracy.

Residential buildings may require multiple stakeouts including grading, building layout, footing, and brick point stakeouts. Commercial buildings and other buildings with atypical foundations or special design characteristics may require unique stakeout methods different from those previously mentioned.

9.9.1. Grading Stakeout for Buildings

Preliminary preparation of a site involves the removal of topsoil and the manipulation of the earth to bring it to subgrade level. To accomplish this, the contractor requires a grading stakeout. The typical stakeout method extends the outermost lines of the building to create a simple rectangle, and the surveyor stakes offsets at the corners. Usually, a 10-foot offset is sufficient to enable earth-moving equipment to move about freely while retaining the stakes for horizontal and vertical reference. Since this stakeout only provides a rough grading guide, the hubs are guarded with tall stakes for easy visibility but not tacked (see Fig. 9.13).

Once the grading stakes are set and checked, the final location will be stored in the data collector for office personnel to verify the stakeout and perform cutsheet calculations. The elevations are marked on the back of each guard. Office personnel check the field notes, and the surveyor issues a cutsheet based on first floor (or finished basement floor, if applicable) grades for the contractor's use.

9.9.2. Construction Stakeout for Buildings

Once the rough grading operation is complete, the building is staked for construction. Based on the grading or site plan, the surveyor establishes the location of the building, and then establishes the corresponding offset hubs. To ensure compliance with zoning setback requirements, the surveyor computes the building location based on the property lines.

Coordinates are set for each corner of the building. Surveyors assign corresponding point numbers and record the points on the building stakeout sheet. Once setting all the building coordinates are set, they compute the building setbacks and side yard distances and confirm all setback distances to property lines comply with local code and restrictions. To maximize lot yield, often the developer will use the minimum setback distances, and therefore, there is little room for movement, or error, in the stakeout of a building. As another precaution, the surveyor can overlay a printout of the building corner stakeout on the site plan to check compatibility.

Ten-foot building stakeout offsets are typically used for building stakeouts as this distance

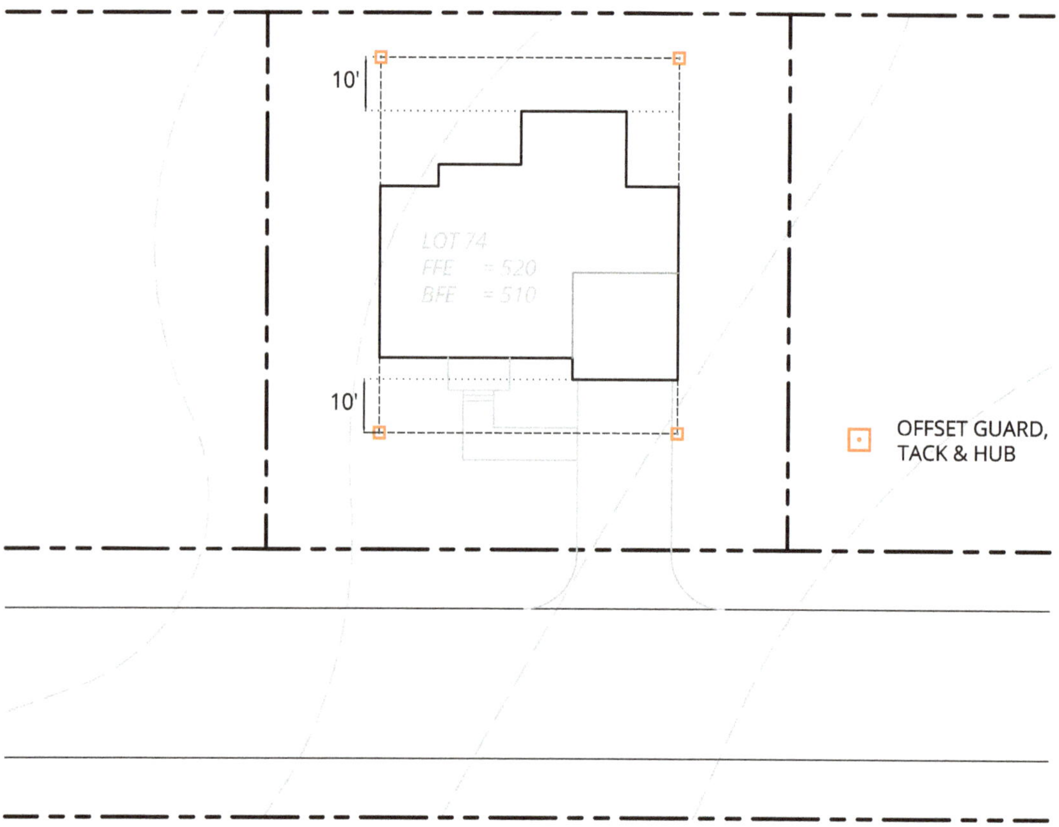

FIGURE 9.13 Building grading stakeout.

is sufficient to accommodate the overexcavation of the basement or foundation and short enough for the contractor to use accurately. The proposed 10-feet offsets are drawn onto the stakeout sheet. The placement is usually from side-to-side, but front-to-back offsets can be used as well. The side-to-side offsets will be on the face of building line 10-feet from the edge of the squared building. The offsets are assigned point numbers and coordinates.

The building stakeout sheet will have coordinate point numbers written next to each lot corner, at intersection points on lot lines, on offset points, and at the building corners. The stakeout sheet may also show control points tied to the offset hubs of the corners of the building. The sheet also includes a list of the point numbers and their coordinates to facilitate any necessary field computations. In addition, the sheet provides the side yard distances and the minimum distances as prescribed by zoning regulations as well as curve information where applicable. The survey crew will file a copy of the building stakeout sheet for future use.

Surveyors determine elevations using at least two benchmarks and establish a front offset hub as a temporary benchmark by using it as a turning point in the level circuit. They use the lowest floor elevation provided on the development plan (either basement or first floor) to calculate the earthwork cut or fill to the finished floor elevation. The hub elevation is written on the stake along with the cut or fill amount. The surveyor checks the field notes and elevations and issues a formal cutsheet.

Footing Stakeout. After excavating the building area, contractors frequently request a footing or foundation stakeout. The surveyor recovers or replaces the building stakeout offset hubs and uses them to lay out the actual building corners. The front (or rear) building offset line may be used as a baseline to transfer the corners to the foundation excavation. Another method is a radial stakeout, where the surveyor occupies one offset and uses the coordinates of each building corner to determine bearings and distances from the setup point. The latter method can be employed successfully if all distances between footing points are checked thoroughly. Diagonal measurements between footing corners check the foundation is square.

The vertical portion of the footing stakeout consists of setting a grade marker, established as a temporary benchmark when used as turning point when the level circuit is run. This point is used as a vertical reference to finish grade.

Brick Points. After the footing is poured and set, the building is ready for a brick point stakeout. This consists of setting concrete nails in the footing at each corner or jog of the building. The masons for block walls or forming crew for poured walls use this stakeout to align the building walls. The stakeout should be very accurate to provide a square foundation for subsequent work.

Vertically, masonry brick points often require grade pegs set to finish floor grade. This involves setting a stake so that the top is equal in elevation to the finish floor grade or an even foot higher or lower, as convenient. Another method uses a stake with a nail driven into its side at grade. Poured wall brick points often circumvent this formality because the foundation is already in place and functions as the critical vertical surface in relation to the finish floor grades. Some surveyors check the elevation of the footer as a precaution for a poured-wall brick point stakeout. A consultation with the developer onsite can determine the necessity of a level run.

Commercial/Retail Buildings. Depending upon the structure's configuration, column lines, baselines, or specified control points control the stakeout for a large building. The key to successful stakeout is to maintain high standards of accuracy and to reference the building control network offsite.

Prior to the stakeout, a traverse control must be set and computed, vertical control established and verified, and architectural plans scrutinized and checked. Ties from the traverse control to the building's control points are computed next, and a stakeout plan is drafted for the survey crew.

As with the other stakeouts described in the preceding sections, the surveyor determines elevations using trigonometric leveling and writes the hub elevations on the stakes. The surveyor issues a formal cutsheet after checking the field notes and the elevations.

9.10. SANITARY SEWER STAKEOUT

Pipe invert elevations dictate the design, construction, and operation of the sanitary sewer. A manhole accommodates horizontal change of direction or vertical grade break in the pipe. Three key elevations for existing or proposed manholes are the top (or rim) elevation, the invert elevation(s) of inflow pipe(s), and the invert elevation of the outflow pipe. A manhole structure may be located according to the centerline of the barrel or the centerline of the manhole cover. For eccentric manholes, the centerline of the cover may be offset 1- or 2-feet from the main barrel centerline, whereas for concentric manholes, the centerlines will be equal. The point used to locate the manhole depends on the purpose and intent of the survey. For as-built surveys and construction purposes, the center of the barrel structure is crucial to pipe slope, pipe length, and easement calculations. For topographic surveys, the center of the manhole cover is often sufficient.

The stakeout proceeds downstream to upstream. Construction typically begins at the lowest point of the stakeout, usually an existing manhole, and progresses upstream. The lowest manhole invert of the sanitary sewer is the beginning with its offset hub called the "out." The beginning manhole is typically designated as station 0+00. In Fig. 9.14, manhole SAN 19 is labeled as OUT 0+00 indicating it is the downstream manhole. Manhole SAN 19 is also given a stakeout number (#1715) corresponding to the cutsheet which provides invert and offset values. If the length of the sanitary sewer line to the next manhole is 100-feet, then the offset station for that manhole is 1+00 and called the "in." As shown in Fig. 9.14, manhole SAN 18 is labeled as IN 1+00 indicating it is the upstream manhole, and its stakeout number is #1738. Another hub is set for this same manhole and designated as the out. This hub is 90° from the bearing of the pipe entering the manhole. At this hub, the stationing reverts to 0+00 again before proceeding to the next manhole. In this manner, the actual distances between manholes are self-evident and can be readily compared to the construction plan, the stakes and the cutsheet pluses.

Intermediate stations are staked and the stakeout length is checked for compatibility to the profile

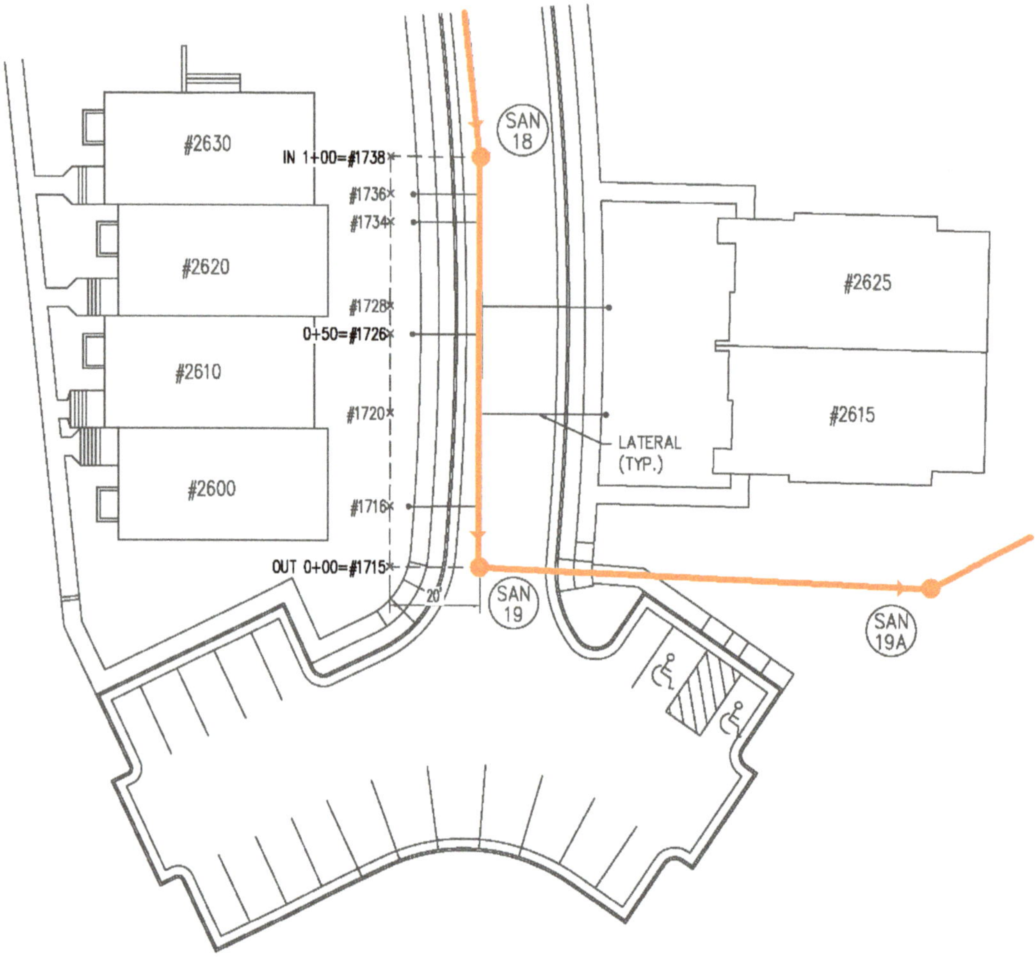

FIGURE 9.14 Typical sanitary sewer stakeout.

(see Fig. 9.14). The offset line stations between the out and the in are generally set every 50-feet, but if the terrain is steep, 25-foot stations may be required to ensure that the pipe crew has enough horizontal and vertical checks as the line progresses. Additional stakeout points are added for the lateral connections to the buildings, as discussed further in the Sanitary Sewer Lateral Stakeout section below.

After the offset hubs are set, the final coordinates of the hub are saved, including elevation data that can be used later for preparing the cut-sheet. The existing manhole where the stakeout began is accurately located with the top rim used as the principal reference point. The existing pipe inverts are measured down from the top elevation. If the proposed sanitary sewer is tying into an existing sanitary between two manholes, both existing manholes must be located vertically and horizontally. The distance between them is measured to determine the slope of the existing pipe. Interpolation is then used to determine the invert pipe elevation of the new manhole.

9.10.1. Sanitary Sewer Lateral Stakeout

The sanitary sewer lateral is the relatively short length of sewer line that runs from the sanitary sewer main to the property line. The lateral's length on the plan view is checked by scale, and the presence of risers (vertical pipes) in the lateral section are noted, as these must also be marked with offset hubs.

For each lateral, a hub is placed at the appropriate station along the sanitary sewer main. The length to the riser, plus the offset distance, is also measured and staked. This method places the stakes on the lateral line extended. An example

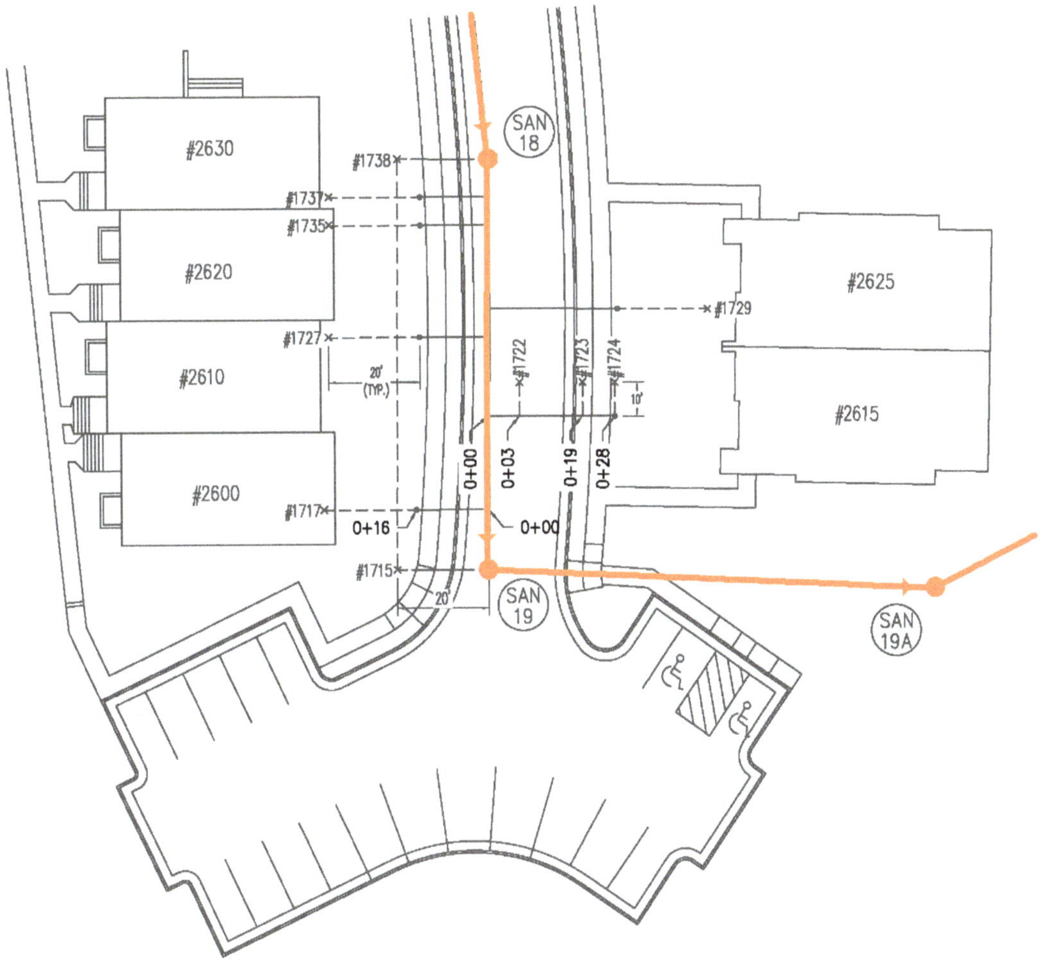

Figure 9.15 Sanitary sewer lateral.

of this method is shown on the lateral for unit #2600 in Fig. 9.15. Another way to offset the lateral is up the sanitary sewer main, at the offset distance. This method is shown for unit #2615 in Fig. 9.15.

9.11. STORM DRAINAGE STAKEOUT

Storm drainage stakeout requires high accuracy as the storm drain structures' tops, grates, and throats must match the road pavement or curb and gutter grade, contours in graded lots, and outfall elevations in existing swales. The surveyor may make corrections in the field and in the cutsheets to ensure positive drainage and eliminate ponding.

9.11.1. Storm Drainage

Street geometry such as stationing, width, and slope determines the location of catch basin. Typically, the storm drainage system is installed before the curb and pavement. The surveyor uses the distance between the centerline of street to the face of curb and face of curb to the centerline of structure to determine the structure's location. The distance from the face of curb to the structure's centerline will vary based on the structure type as shown in the detail drawings.

Many contractors request stakes relating the storm drainage structure to the face of curb. This ensures accurate placement of the structures to match the curb and gutter installation. Along tangent sections, offsets are set along the face of curb with the offset distances relating to the centerline of structure indicated on the stakes. In a curve, the offsets are placed on a line perpendicular to the radial line. Two hubs will typically be used for a catch basin (or even three for larger catch basins)

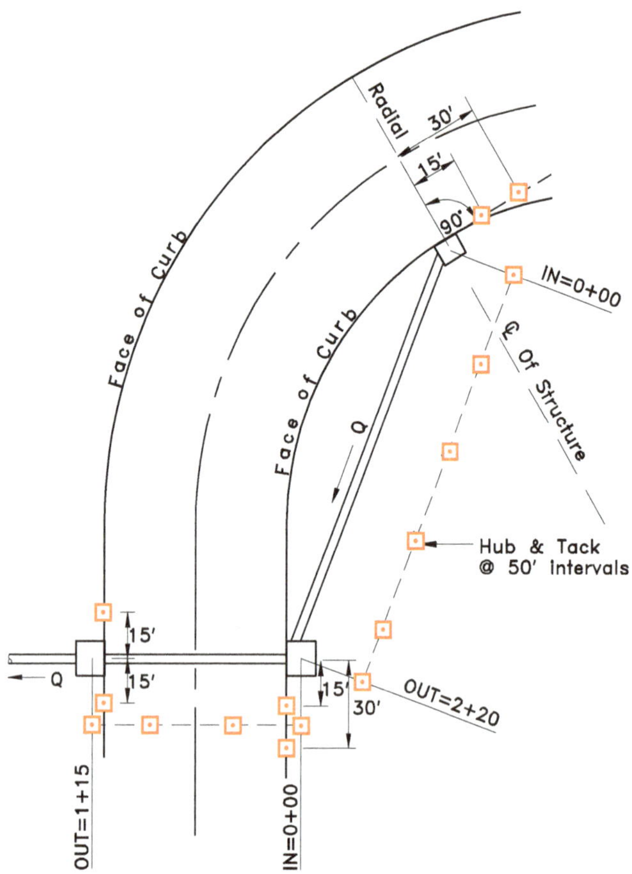

Figure 9.16 Face of curb offsets for storm drainage stakeout.

and one hub is typically sufficient for a drop inlet. Both of these methods are shown in Fig. 9.16.

The storm pipe offsets are staked once the centerline of each structure and the face of curb (if applicable) hubs are set. The offset distance is determined with the contractor and by the depth of pipe shown in the plans and profiles. Onsite conditions and practical considerations usually determine the placement of the offsets on one side of the pipe. For example, the surveyor will typically place the offset on the side opposite from the street, on the side most protected from construction machinery, and/or the "high side" of the construction area.

The hub for the pipe that discharges into the downstream structure is designated as the OUT, with a station 0+00 (see Fig. 9.17). The IN hub is the inlet side of the same pipe, at the next upstream structure, whose station is designated by the distance between the two structures. Offset hubs along the pipe length are set at 50-foot intervals, while any pipe and utility crossings are offset as well. The profile in the site plans provides stations for any crossings. Crossings staked in the field help the contractor avoid contact with existing utilities by maintaining the minimum amount of clearance. Elevations are determined using a total station instrument.

9.11.2. Stormwater Management Ponds

The surveyor stakes stormwater management ponds along the proposed grading contours. Hubs are typically set to locate the toe of the proposed berm and to mark the top of the proposed berm. Stakeout markers are set at horizontal bends and at regular intervals to delineate the size and shape of the pond. However, setting too many markers may conflict with the operation of large earth-moving equipment. Therefore, the surveyor typically stakes the proposed contours every 2- to 3-feet.

Preliminary preparation for this stakeout includes plotting traverse or other control points on the site plan with computed coordinates for each point. A total station instrument and data collector may be used to stake the points, which usually consist of a hub and a 4-foot tall guard stake. The hub represents the horizontal and vertical point. The guard conveys the necessary information (written directly on the stake itself) such as the description (proposed contour, top of bank, or toe of bank), the hub elevation, and the cut or fill information. Traditionally, a pond stakeout required minimum accuracy, typically to the nearest foot horizontally and to the nearest tenth of a foot vertically (see Fig. 9.18). However, with increasingly stringent stormwater regulations and implementation of Low Impact Development (LID) stormwater practices such as bioretention and dry swales, a higher degree of accuracy is required for the facility's construction and therefore, the original stakeout. LID stormwater facilities often have subsurface layers for additional stormwater volume retention and treatment in addition to the at-grade basin. Jurisdictions will require as-built surveys to demonstrate these constructed stormwater facilities meet or exceed the calculated volume required for stormwater quality and/or quantity treatment.

Dewberry
8401 Arlington Blvd
Fairfax VA 22031
703-849-0100

Project: Mt Vernon Hospital
Job #: 1202-0430

Storm Sewer Cutsheet

Grade To: Invert of Pipe or As Noted
Location: Strc 30-31-32-33-34-35, 32-36-37

Prepared by: CAP **Date:** 04/13/2025
Staked by: SMP **Date:** 04/13/2025
Checked by: LHP **Date:** 04/13/2025

Description	Offset	Station	Stake Elev	Grade Elev	Cut/Fill	Comments
__SMSO36__						
STRC 32(IN) MH	15' RT	00+00	261.80	259.30	C-2.50	204.01- 15" RCP @ 1.57%
	15' RT	00+50	263.82	260.08	C-3.74	
	15' RT	01+00	264.48	260.87	C-3.61	
	15' RT	01+50	265.96	261.65	C-4.31	
STRC 36(OUT) INLET	15' RT	02+04	267.57	262.50	C-5.07	
FC @ STRC 36	15' LT	13+73	267.14	269.02	F+1.88	TOP OF CURB
FC @ STRC 32	30' LT	13+88	267.44	269.02	F+1.58	TOP OF CURB
__SMSO37__						
STRC 36(IN) INLET	15' RT	00+00	267.10	262.60	C-4.50	32.00' - 15" RCP @ 0.62%
	15' RT	00+16	267.16	262.70	C-4.46	
STRC 37(OUT) INLET	15' RT	00+32	267.04	262.80	C-4.24	
FC @ STRC 37	15' RT	13+73	267.08	269.02	F+1.94	TOP OF CURB
FC @ STRC 37	30' RT	13+88	267.34	269.02	F+1.68	TOP OF CURB

FIGURE 9.17 Storm drainage cutsheet.

9.12. WATERLINE STAKEOUT

Waterlines located within the right-of-way are typically parallel to the street centerline and use the same street stationing. Waterline joints allow for some deflection in the alignment, and fittings such as horizontal or vertical bends provide for abrupt changes in direction. Prior to commencing the waterline stakeout, the surveyor will verify the information on the plan and profile including the location of all valves, fittings, and fire hydrants.

When a waterline is located within an easement through a private site, it will have its own stationing system as shown in the plan and profile. The surveyor will follow this stationing as closely as possible as this is the system approved by the local jurisdiction. If there are inaccuracies, a surveyor may provide a station equality on the stakes with the plan and profile station noted first, followed by the equivalent stakeout station. This allows for correction in the stationing to match the remainder of the stakeout to the plan stationing.

9.12.1. Fire Hydrants

Fire hydrants must be accurately placed horizontally and vertically. Fire protection requirements typically establish the location of a hydrant based on the fire department's procedures during a fire event.

The stakeout is based on the relationship between the curb line, edge of pavement, or shoulder and the main waterline. The distance from street centerline to the face of curb is the reference point that controls the hydrant location. As shown in Fig. 9.19, a hub marking the tee for the stub line leading to the fire hydrant is set along the offset line for the main waterline. Two offset hubs are set beyond the fire hydrant a minimum of 10-feet apart and along the alignment of the fire hydrant line. Two

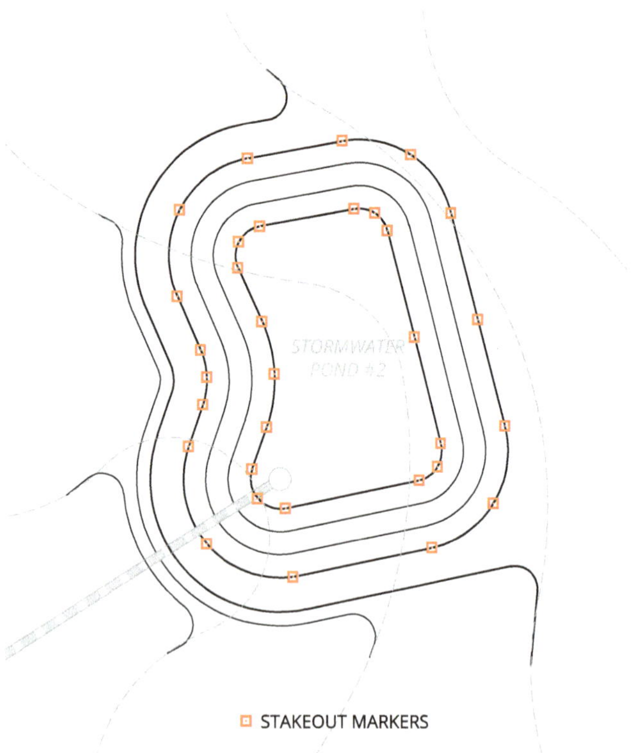

Figure 9.18 Pond contour and stakeout points.

additional hubs are then set marking the edge of shoulder (alternatively, these hubs could mark the curb line or edge of pavement). A string line is stretched between the hubs marking the shoulder,

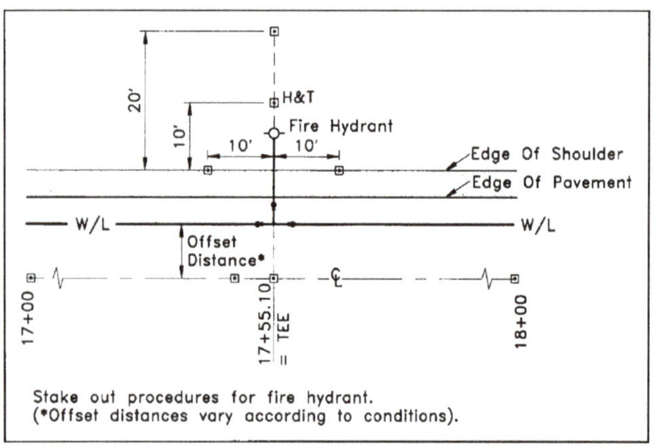

Figure 9.19 Stakeout procedures for fire hydrant.

and additional strings tie the hubs beyond the fire hydrant to the hub for the tee. The intersection of these strings mark the intersection of the shoulder and the waterline stub to the fire hydrant. A guard stake is marked to indicate the elevation of the shoulder and the contractor can install the fire hydrant based on the guard stake.

9.12.2. Service Lines

A separate stakeout is prepared for water meter boxes, where public service ends and a private line extends to serve the building. Meter boxes contain the water meter and cutoff valves for each service line. The meter box location is marked either with paint on the curb or with hubs and guards. The local jurisdiction typically stipulates the distance between the property line and the meter box. Either the property line is marked at the proposed meter box location or the property line is referenced by offsets in order to preserve the stakeout. The water meter box can also be staked in line with the stub, just beyond the property line. One offset with an elevation given for each meter box is generally adequate to build the meter box and its lateral from the main.

Waterline laterals that serve high-rise office and apartment buildings are staked with multiple offsets, indicating the tee location, intermediary station(s), bends, and the connection to the building. The building connection location may differ from the site plan, as architectural plans are not always finalized before the site plans are approved. A consultation with the contractor or plumber is recommended before staking out building connections. At times, dimensions pinpointing the plumbing inside the building are measured and recreated outside to locate the connecting point. The stakeout then proceeds in the above-described fashion. Fig. 9.20 shows an example waterline cutsheet.

Dewberry
10003 Derekwood Lane
Lanham, Maryland 20706-4804
ph.301 731 5551 fax.301 731 0188 www.dewberry.com

CONTRACT #____________ **CS#021**
SHEET# __1___ OF ___1___
REVISION DATE __________

Date: 1/9/07 | Job: RIDGES 2&3 | BENCHMARK:

LOCATION: *WOODREED DRIVE* | PARTY CHIEF: DV

DESCRIPTION: **WATERLINE STAKE OUT** | PREPARED: DSD

Point	Offset	Station	Hub Elev	Prop Elev	Cut	Fill	Remarks
WATER LINE STAKEOUT WOODREED DRIVE							
1516	10' RT	0+00	TEST PIT	215.30	TEST PIT		OS10 WL (TEST PIT)
1517	10' RT	0+15.85	220.91	215.11	5.80		OS10 WL
1518	10' RT	0+50	220.79	214.70	6.09		OS10 WL
1519	10' RT	1+00	219.31	213.50	5.81		OS10 WL
1520	10' RT	1+50	216.07	211.60	4.47		OS10 WL
1521	10' RT	2+00	212.92	208.85	4.07		OS10 WL
1522	10' RT	2+50	208.81	206.10	2.71		OS10 WL
1523	10' RT	3+00	204.76	202.30	2.46		OS10 WL
1524	10' RT	3+50	200.37	198.10	2.27		OS10 WL
1525	10' RT	4+00	195.85	193.20	2.65		OS10 WL
1527	10' RT	4+24.51	193.25	190.75	2.50		OS10 WL
1528	10' RT	4+50	190.81	188.20	2.61		OS10 WL
1529	10' RT	4+95	186.39	182.70	3.69		OS10 WL
1530	10' RT	5+25	182.99	180.50	2.49		OS10 WL
1531	10' RT	5+64	178.87	176.48	2.39		OS10 WL
1546	0	5+64	179.03	182.07		3.04	CL FH (EDGE PAVING EL.)
1545	10'	5+64	178.63	182.07		3.44	OS10 FH (EDGE PAVING EL.)
1532	10'RT	6+07/0+00	174.45	171.90	2.55		OS10 WL
1534	10'RT	6+17.99	173.26	170.73	2.53		OS10 WL@ WV
1535	10'RT	6+50	169.70	168.50	1.20		OS10 WL
1536	10'RT	7+00	164.27	163.50	0.77		OS10 WL
1537	10'RT	7+50	159.22	158.50	0.72		OS10 WL
1538	10'RT	7+95.32	154.95	153.97	0.98		OS10 WL
1539	10'RT	8+00	154.58	153.50	1.08		OS10 WL
1540	10'RT	8+50	150.80	148.50	2.30		OS10 WL
1541	10'RT	9+00	147.37	143.50	3.87		OS10 WL
1542	10'RT	9+40	141.68	139.50	2.18		OS10 WL
WATER HOUSE CONNECTIONS							
1543	10	LOT 9	169.46	165.00	4.46		FINAL GRADE
1544	10	LOT 4	165.83	172.00		6.17	FINAL GRADE

NOTE: SURVEYOR DOES NOT TAKE THE RESPONSIBILITY FOR ANY ALLEGED ERROR OR DISCREPANCY FOUND IN THE FIELD OR ON THIS CUT SHEET UNLESS DEWBERRY IS NOTIFIED IMMEDIATELY SO THAT THE MATTER CAN BE CHECKED AND RESOLVED. STAKES PLACED IN THE FIELD BY DEWBERRY AND DAVIS MUST BE UNDISTURBED AND IN PLACE SO THEY CAN BE FIELD CHECKED AND VERIFIED. CONTRACTORS MUST CHECK THREE OR MORE HUB ELEVATIONS PRIOR TO THE START OF CONSTRUCTION.

Figure 9.20 Waterline cutsheet.

CASE STUDY: UNIDENTIFIED UTILITIES

ISSUE:

Contractual allocation of risk regarding underground facilities.

SUMMARY:

Five million municipal sewer project. Best Construction, the contractor, brought a breach of contract claim against the owner, alleging that during construction the contractor encountered 122 unmarked or mismarked underground utilities, resulting in delay and additional costs, and also contending that the owner allowed other contractors to conduct work that interfered with Best Construction's work.

At a preliminary stage in the lawsuit, the owner moved for dismissal based on lack of written notice of claim, and on the argument that the construction contract allocated the risk of underground facilities to the contractor. The trial court granted the motion to dismiss, and the contractor appealed.

DECISION:

The court of appeals allowed the contractor's case to continue, reversing the lower court's dismissal.

The construction contract contained an "Underground Facilities" clause very similar to the provisions of what is currently Paragraph 5.05 of the EJCDC Standard General Conditions (C-700). The appellate court concluded that the trial court had erroneously ignored the terms of this clause, which allows for an adjustment in contract time and contract price if the contractor encounters an underground facility that is not shown on the drawings, or not shown accurately.

The appellate court also determined that although the contractor had not filed timely formal notices, it was possible that the owner had received adequate actual notice of the contractor's underground facilities claims. The pleadings alleged that the owner's representative had documented each subsurface incident, and that the owner had waived strict compliance with many other procedural provisions of the contract. If proven, such allegations might be sufficient to overcome the lack of formal notice.

COMMENT:

The EJCDC Underground Facilities clause makes clear that the contractor is entitled to additional compensation and time for the cost of contending with unmarked and mismarked utilities at the jobsite. However, it is possible that a reader unfamiliar with the Underground Facilities clause might focus on the parts of the clause that emphasize the contractor's responsibilities for the safety and protection of all utilities—but that responsibility is not uncompensated.

According to the somewhat abbreviated written decision, the contractor did not begin to pursue its claim until after receipt of final payment. Most construction contracts, including EJCDC® C-700, indicate that final payment closes the opportunity for contractor to make claims. The written decision does not mention whether this sewer contract contained such a clause, or whether there were factors that allowed for post-payment claims.

Case study based on J.F. Allen Corporation v. Sanitary Board of City of Charleston, West Virginia. Supreme Court of Appeals of West Virginia (2016).

The information provided does not constitute legal advice. Consultation with an attorney is strongly advised.

9.13. CONCLUSION

Construction stakeout requires substantial preparation before sending the survey crew to the field. Successful computing requires familiarity with construction techniques as well as good surveying skills. Inevitably, there will be times when the survey crew will have to adjust the stakeout procedure in the field due to unforeseen job site conditions. Therefore, the field crew must take all of the plans, plats, and hardcopies as well as geometric coordinates with them in order to perform calculations necessary to complete the task. Delays due to poorly prepared survey crews can delay the project and ultimately increase construction costs.

The procedures for the various stakeout methods described above are intended as a suggestion of the ways to perform construction stakeouts. Many localities have regulations that dictate construction methods and preferences for performing construction stakeout due to local conditions and biases. In addition, many surveyors have adopted other ways to perform stakeout work due to the type of field and office instruments available, local contractors' preferences, and field conditions. Before employing any of the preceding methods, it is best to consult local ordinances and public inspectors to avoid conflicts and ensure compliance with the prevailing regulations.

Appendix A

Construction Equipment

FIGURE A.1 ***Excavator.*** Excavators are manufactured in a variety of sizes for applications in large and small construction projects. They rely upon a hydraulic system to operate the machine's mechanical arm. The mechanical arm is used for various activities such as trenching, debris removal, demolition, material lifting/placement, and general grading.

FIGURE A.2 ***Asphalt compactor.*** Asphalt compactors are also referred to as asphalt rollers or tandem rollers. They can be manufactured with two (tandem) or three drums and compact asphalt by means of weight (static) or vibration (vibratory). Following the asphalt surface course placement, asphalt compactors reduce voids throughout the pavement section to improve strength and minimize failure potential.

FIGURE A.3 ***Soil compactor.*** Soil compactors, or rollers, are used to increase soil strength by removing voids and densifying the subsurface. They can be static or vibratory, and the drum can be a smooth or sheepsfoot configuration. Sheepsfoot compactors are typically used for fine-grained, clay soils. Soil compactors are used where relatively higher compaction percentages are desired at greater depths such as with roadway and railway construction.

FIGURE A.4 ***Asphalt paver.*** Asphalt pavers consist of a hopper, auger, and screed which are used to process, place, and provide minor compaction of hot mix asphalt. The screed is used to control asphalt placement to achieve the proposed elevations and pavement sections.

FIGURE A.5 ***Backhoe loader.*** A backhoe loader is a combination between a tractor, loader, and backhoe. The tractor component allows the machine to maneuver uneven terrain with oversized, durable tires. The loader is typically used to collect or plow loose material. The backhoe is designed to excavate compacted material or lift heavy loads such as concrete pipes or sewer structures.

FIGURE A.6 ***Skidder.*** Skidders are used to clear vegetation from a site using the push blade on the front to demolish and the winch on the back side to collect the fallen debris. They can be used to selectively demolish smaller vegetation while retaining the larger, mature vegetation for final conditions.

FIGURE A.7 ***Front loader***. Front loaders are used to collect loose material such as gravel, soil, or debris in an elevated manner without pushing material from initial to ultimate location. They are not typically used for grading, as they provide minimal excavation below their wheel level.

FIGURE A.8 ***Dozer.*** A dozer, or bulldozer, is equipped with a blade on the front end and sometimes a ripper on the rear end. The ripper is designed to loosen compacted material such as soil or aggregate, and the blade is designed to push loose material. The blade can vary in shape depending on the function. The dozer's tracks allow it to maneuver through uneven terrain.

FIGURE A.9 ***Pipelayer.*** Pipelayers are used to lower individual pipe sections into a trench where construction crews are positioned to set the pipe. They use tracking maneuverability to move slowly along a trench as they lay pipe.

FIGURE A.10 ***Horizontal drill.*** Horizontal drills are used in trenchless pipe installation. They typically drill a pilot hole beneath and perpendicular to linear features such as roadways, rivers, and wetlands, and then an enlarged hole is created to allow for the diameter of the conduit to be installed.

FIGURE A.11 ***Grader.*** Graders (also known as a road grader or motor grader) are used for a variety of grading operations. They can be used to fine grade a site, establish base course elevation for a roadway, or subgrade elevation for a structural element such as a building pad.

FIGURE A.12 ***Skid steer loader.*** Skid steer loaders, or skid steers, are small-scale front loaders that typically maintain a zero turn radius. These are particularly useful on smaller sites such as single-family residential lots. Skid steers are primarily used to transport loose material, but can also be used for minor and fine-grading activities in small-scale applications.

FIGURE A.13 ***Pile hammer.*** A pile hammer, or pile driver, is a mechanism that drops a weight on a building pile foundation. It uses a crane mechanism to raise a piston that is dropped on the pile to drive the pile to the appropriate sub-grade elevation after a sufficient number of blows.

FIGURE A.14 ***Crane.*** Cranes are manufactured in a variety of styles for various uses. Styles include vehicle mounted, tower, rough terrain, crawler, all terrain, railroad, telescope handler, harbor, floating, aerial, telescopic, and level luffing. They are generally used for elevating material from ground level to elevated construction surfaces.

FIGURE A.15 ***Dump truck.*** Dump trucks are designed to receive and haul loose material such as aggregate, soil, and other construction debris onsite as well as to offsite disposal facilities. They often operate in tandem with front loaders to move loose material.

FIGURE A.16 ***Fork lift crane.*** A fork lift crane is designed to lift pallets or other objects that can be lifted on the flat surface created by the two forks located at the front end of the crane. Similar to other cranes, it maintains a load capacity proportional to the counterweight provided along the rear end of the crane.

FIGURE A.17 ***Hoe ram.*** A hoe ram, or breaker, is a hydraulically operated hammer device that can be fitted to an excavator to demolish rigid structures such as pavement or rock. It is typically used on larger projects where jack hammers are inefficient but also where blasting is inappropriate due to the surrounding environment.

APPENDIX B

SPECIFICATIONS

MASTERSPEC building specs done right

TABLE OF CONTENTS
Site Civil Library

Section No.	Section Title	Section Description
DIVISION 00 - PROCUREMENT AND CONTRACTING REQUIREMENTS		
000101	Project Title Page	Project Manual title page.
000107	Seals Page	For seals of design professionals on Project Manual.
000115	List of Drawing Sheets	List of Drawings.
000120	List of Schedules	For a list of separately bound schedules.
001113	Advertisement for Bids	For use on public projects.
001115	Advertisement for Prequalification of Bidders	For use on public projects.
001116	Invitation to Bid	For use on private projects.
001153	Request for Qualifications	For use on public projects.
002113	Instructions to Bidders	Incorporates AIA A701 by reference.
002213	Supplementary Instructions to Bidders	For use with AIA A701.
002513	Prebid Meetings	Sets date, time, place, and terms for prebid meetings.
002600	Procurement Substitution Procedures	Substitution procedures during bidding.
003113	Preliminary Schedules	Preliminary schedules.
003119	Existing Condition Information	References documents for survey and as-built information.
003126	Existing Hazardous Material Information	References documents for existing hazardous materials.
003132	Geotechnical Data	References documents for geotechnical data.
003143	Permit Application	Indicates responsibility for building permit application.
004113	Bid Form - Stipulated Sum (Single-Prime Contract)	Fixed price, single prime.
004116	Bid Form - Stipulated Sum (Multiple-Prime Contract)	Fixed price, multiple prime.
004123	Bid Form - Construction Management (Single-Prime Contract)	Fixed price, construction management, single prime.
004126	Bid Form - Construction Management (Multiple-Prime Contract)	Fixed price, construction management, multiple prime.
004133	Bid Form - Cost-Plus-Fee (Single-Prime Contract)	Cost plus, with or without guaranteed maximum.
004313	Bid Security Forms	References AIA A310.
004321	Allowance Form	Attachment for Bid Form.
004322	Unit Prices Form	Attachment for Bid Form.
004323	Alternates Form	Attachment for Bid Form.
004373	Proposed Schedule of Values Form	Attachment for Bid Form.
004393	Bid Submittal Checklist	Attachment for Bid Form.
005100	Notice of Award	For notifying successful bidder.
006000	Project Forms	References Agreement, General Conditions, and administrative forms.
009113	Addenda	For notifying bidders of changes.
DIVISION 01 - GENERAL REQUIREMENTS		
010000	General Requirements	Evaluations: Role of Division 1 Sections, their preparation and coordination.
011000	Summary	Summary of the Work, phased construction, purchase contracts, Owner-furnished products, access to site, and work restrictions.
011200	Multiple Contract Summary	Responsibilities of each contract for the Work, coordination, and temporary facilities and controls.
012100	Allowances	Provisions for cash allowances including lump-sum, unit cost, contingency, and testing and inspecting allowances.
012200	Unit Prices	Provisions for unit prices.
012300	Alternates	Provisions for change-of-scope and cost-comparison type alternates.

N = New U = Update

FIGURE B.1 MasterSpec's Site Civil Library table of contents (provided courtesy of Avitru, the developer of AIA MasterSpec).

MASTERSPEC building specs done right

TABLE OF CONTENTS
Site Civil Library

Section No.	Section Title	Section Description
012500	Substitution Procedures	Procedural requirements for requests for substitutions during construction.
012600	Contract Modification Procedures	Procedural requirements for changes to the Contract.
012900	Payment Procedures	Administrative requirements for Contractor's Applications for Payment.
013100	Project Management and Coordination	Administrative requirements for project meetings; preconstruction, preinstallation, and project closeout conferences; RFIs; and project Web sites.
013200	Construction Progress Documentation	Contractor's Construction Schedule including Gantt charts and CPM schedules; and web-based project software.
013233	Photographic Documentation	Construction photographs, video recordings, and web-based photographic documentation.
013300	Submittal Procedures	Procedures for Action and Informational Submittals including Delegated-Design Submittals and Submittals Schedule.
013516	Alteration Project Procedures	General protection and work procedures for remodeling, renovation, repair, and maintenance work.
014000	Quality Requirements	Quality-assurance and -control requirements, special tests and inspections, and Contractor's quality-control plan.
014200	References	Common definitions and terms; and acronyms and trade names of associations, government agencies, and other entities referenced in MasterSpec.
015000	Temporary Facilities and Controls	Temporary utilities and facilities for construction support, security, and protection.
015639	Temporary Tree and Plant Protection	Construction-phase tree and plant protection, trimming, protection- zone fencing, and pruning.
015723	Temporary Storm Water Pollution Control	Requirements for compliance with EPA's Stormwater Pollution Prevention Plan (SWPPP).
016000	Product Requirements	Administrative and procedural requirements for product, material, and equipment selection and handling; warranties; and comparable products.
017300	Execution	General requirements for product installation, cutting and patching, protection, field engineering, and progress cleaning.
017419	Construction Waste Management and Disposal	Salvaging, recycling, and disposing of non-hazardous demolition and construction waste.
017700	Closeout Procedures	Contract closeout including Substantial Completion and Final Completion procedures, warranties, and final cleaning.
017823	Operation and Maintenance Data	Emergency, operation, and maintenance manuals for products and equipment.
017839	Project Record Documents	Record record Drawings, Specifications, and Product Data.
017900	Demonstration and Training	Administrative and procedural requirements for instructing Owner's personnel in operation and maintenance.
018113.13	Sustainable Design Requirements - LEED 2009 for New Construction and Major Renovations	General requirements and procedures for LEED 2009 for New Construction and Major Renovations.
018113.14	Sustainable Design Requirements - LEED v4 BD+C	General requirements and procedures for LEED v4 BD+C.
018113.16	Sustainable Design Requirements - LEED 2009 for Commercial Interiors	General requirements and procedures for LEED 2009 for Commercial Interiors.
018113.17	Sustainable Design Requirements - LEED v4 ID+C	General requirements and procedures for LEED v4 ID+C.
018113.19	Sustainable Design Requirements - LEED 2009 for Core and Shell Development	General requirements and procedures for LEED 2009 for Core and Shell Development.
018113.23	Sustainable Design Requirements - LEED 2009 for Schools	General requirements and procedures for LEED 2009 for Schools.
018113.33	Sustainable Design Requirements - IgCC	General requirements and procedures for IgCC.

N = New U = Update

AVITRU

Figure B.1 *(Continued)*

MASTERSPEC building specs done right

TABLE OF CONTENTS
Site Civil Library

Section No.	Section Title	Section Description
018113.43	Sustainable Design Requirements - ASHRAE 189.1	General requirements and procedures for ASHRAE 189.1.
018113.53	Sustainable Design Requirements - Green Globes	General requirements and procedures for Green Globes.

DIVISION 02 - EXISTING CONDITIONS

Section No.	Section Title	Section Description
024116	Structure Demolition	Demolition of existing buildings, structures, and associated site improvements.
024119	Selective Demolition	Demolition of selected portions of existing buildings, structures, and associated site improvements.

DIVISION 03 - CONCRETE

Section No.	Section Title	Section Description
030130	Maintenance of Cast-In-Place Concrete	Concrete patching and repair, epoxy crack injection, corrosion inhibiting surface treatment, polymer overlays and sealers, and composite structural performance.
033816	Unbonded Post-Tensioned Concrete	Post-tensioning reinforcement, encapsulated and non-encapsulated.
U 034100	Precast Structural Concrete	Conventional precast structural concrete units.

DIVISION 05 - METALS

Section No.	Section Title	Section Description
055000	Metal Fabrications	Metal items (not sheet metal) made from iron and steel shapes, stainless steel, and non-ferrous metals.
055213	Pipe and Tube Railings	Railings fabricated from aluminum, stainless-steel, and steel pipe and tubing.

DIVISION 07 - THERMAL AND MOISTURE PROTECTION

Section No.	Section Title	Section Description
071113	Bituminous Dampproofing	Cold-applied, cut-back- (solvent-based) and emulsified- (water-based) asphaltic dampproofing.
071326	Self-Adhering Sheet Waterproofing	Self-adhering, positive-side sheet waterproofing; drainage panels; pedestal-supported concrete pavers.
071353	Elastomeric Sheet Waterproofing	Butyl and EPDM sheet; drainage panels; pedestal-supported concrete pavers.
071354	Thermoplastic Sheet Waterproofing	PVC sheet; drainage panels; pedestal-supported concrete pavers.
071413	Hot Fluid-Applied Rubberized Asphalt Waterproofing	Hot, rubberized-asphalt waterproofing for plaza deck and below-grade walls.
071416	Cold Fluid-Applied Waterproofing	Polyurethane, polyester, and rubber types; drainage panels and plaza-deck pavers.

DIVISION 11 - EQUIPMENT

Section No.	Section Title	Section Description
110513	Common Motor Requirements for Equipment	Universal default values for small and medium, ac motors.

DIVISION 22 - PLUMBING

Section No.	Section Title	Section Description
U 221113	Facility Water Distribution Piping	Utility service, domestic, and fire-protection water, specialties, and valves.
U 221216	Facility Elevated, Potable-Water Storage Tanks	Above-ground water storage.
U 221219	Facility Ground-Mounted, Potable-Water Storage Tanks	Surface-mounted water storage.
221313	Facility Sanitary Sewers	Sanitary sewerage and underground structures outside the building.
U 221323	Sanitary Waste Interceptors	Grease, oil, and sand interceptors for sewerage systems outside the building.
U 221343	Facility Packaged Sewage Pumping Stations	Pump, compressor, tank, and controls.
U 221353	Facility Septic Tanks	Tank, distribution box, and drainage pipe.

N = New U = Update

AVITRU

FIGURE B.1 *(Continued)*

MASTERSPEC
building specs done right

TABLE OF CONTENTS
Site Civil Library

Section No.	Section Title	Section Description
221413	Facility Storm Drainage Piping	Stormwater piping within the building.
221423	Storm Drainage Piping Specialties	Roof drains, cleanouts, trench drains, channel drains, and piping specialties.

DIVISION 23 - HEATING, VENTILATING, AND AIR CONDITIONING (HVAC)

Section No.	Section Title	Section Description
U 231123	Facility Natural-Gas Piping	Natural gas piping and specialties.
U 231126	Facility Liquefied-Petroleum Gas Piping	LP gas piping, specialties, and storage equipment.
232113.33	Ground-Loop Heat-Pump Piping	Horizontal and vertical, direct-buried, ground-loop piping.

DIVISION 26 - ELECTRICAL

Section No.	Section Title	Section Description
260543	Underground Ducts and Raceways for Electrical Systems	Manholes, handholes, and underground ducts.

DIVISION 31 - EARTHWORK

Section No.	Section Title	Section Description
311000	Site Clearing	Clearing, grubbing, vegetation protection, stripping and stockpiling topsoil and rock, and disconnecting utilities.
312000	Earth Moving	Excavating, filling and backfilling, compacting, and grading.
312319	Dewatering	Temporary construction dewatering requirements for ground-water control.
315000	Excavation Support and Protection	Temporary systems; soldier piles and lagging, sheet piling, tiebacks, and bracing.

DIVISION 32 - EXTERIOR IMPROVEMENTS

Section No.	Section Title	Section Description
U 321216	Asphalt Paving	Paving, overlays, surface treatments, asphalt curbs, asphalt traffic-calming devices, cold milling, and hot-mix patching.
U 321223	Imprinted Asphalt	Embossing patterns in new or existing asphalt pavement.
321313	Concrete Paving	Concrete curbs and gutters, walks, driveways, parking lots, and miscellaneous paving.
321316	Decorative Concrete Paving	Surface-imprinted, stamped, stencil-patterned, or stained finish concrete paving.
321373	Concrete Paving Joint Sealants	Sealants for concrete and asphalt pavement joints.
321400	Unit Paving	Brick, asphalt, concrete, and stone pavers on aggregate, mortar, and bituminous setting beds.
321443	Porous Unit Paving	Concrete, grid-type and interlocking type with openings between units.
U 321713	Parking Bumpers	Wheel stops applied to asphalt or concrete pavement.
U 321716	Manufactured Traffic-Calming Devices	Preformed rubber or plastic traffic-calming devices applied to asphalt or concrete pavement.
U 321723	Pavement Markings	Painted markings applied to asphalt or concrete pavement.
321726	Tactile Warning Surfacing	Detectable warning tiles and mats located in concrete walkways; detectable warning pavers.
323113	Chain Link Fences and Gates	Standard chain-link fencing systems with swing or horizontal-slide gates and gate operators.
323113.53	High-Security Chain Link Fences and Gates	High-security chain-link fencing systems with swing or horizontal-slide gates and gate operators.
323223	Segmental Retaining Walls	Dry-laid concrete masonry unit retaining walls.
328400	Planting Irrigation	Heads, pipes, and controls.
329113	Soil Preparation	Planting soils specified according to quantities of amendments (compost, lime, sulfur, fertilizer, etc.); layered soil assemblies.

N = New U = Update

AVITRU

FIGURE B.1 (Continued)

MASTERSPEC building specs done right

TABLE OF CONTENTS
Site Civil Library

Section No.	Section Title	Section Description
329115	Soil Preparation (Performance Specification)	Planting soils specified according to tested performance requirements (particle size, organic matter percentage, pH, CEC, fertility, etc.); layered soil assemblies.
329200	Turf and Grasses	Seeded, sodded, plugged, and sprigged turf and meadows; pesticides; erosion-control materials; turf renovation; and grass paving.
329300	Plants	Nursery-grown trees and other plants, pesticides, tree stabilization, tree watering devices, landscape edgings, and tree grates.
329600	Transplanting	Transplanting non-nursery-grown trees; tree stabilization; watering devices.

DIVISION 33 - UTILITIES

Section No.	Section Title	Section Description
330500	Common Work Results for Utilities	Basic piping materials and methods.
334200	Stormwater Conveyance	Gravity and forced storm piping, underground structures, and specialties.
334600	Subdrainage	Drainage for foundations, underslabs, plaza decks, retaining walls, and landscaped areas.
334713	Pond and Reservoir Liners	Geomembrane liners and covers.

FIGURE B.1 *(Continued)*

SECTION 01 33 00 - SUBMITTAL PROCEDURES

PART 1 - GENERAL

1.1 SUBMITTAL FORM AND LOG

A. GC shall use either the Submittal Register available through the University's website one of its own making acceptable to the University.

B. GC shall maintain in the field office a copy of the Submittal Log/Schedule indicating the current status of each item.

D. GC shall allot time in the construction scheduling for liaison with the design team for review of Submittals. Except for special Submittals, such as those with design calculations and/or detailed Shop Drawings, Submittals will be reviewed in an average time of 10 working days. Submittal items that are inter-related should be submitted at the same time.

E. Schedule finish sample Submittals for approval of color, texture, graining, or other finish at least 30 days before purchase, assembly, or fabrication.

1.2 SUBMITTALS, GENERAL

A. Design team will return Submittals without processing if they do not contain the GC's stamp, if they are not initialed or signed by an authorized person, if they are not dated, or if it becomes evident that they have not been properly reviewed. Resultant delays are the responsibility of the GC.

B. GC shall bind Submittals in a manner suitable for 8-1/2 × 11-inch file folder storage, except where doing so is not workable.

C. GC shall transmit Submittals with all transportation charges prepaid.

1.3 SHOP DRAWINGS

A. GC shall bind in complete sets and provide digital copies. Folding of prints is acceptable.

B. GC shall allot time in construction scheduling for liaison with the design team for review of Shop Drawings. Except for special Submittals, such as those with design calculations and/or detailed Shop Drawings, Submittals will be reviewed in an average time of 10 working days.

C. GC shall present all drawings in a clear and consistent manner. Each drawing shall contain the project name and number; identify each element of the drawings by reference to sheet number and detail, schedule, or room number of the Contract Documents.

D. GC shall show detail, materials, dimensions, thickness, methods of assembly, attachments, relation to adjoining Work, and other pertinent data and information.

E. GC shall verify dimensions and field conditions and clearly indicate both.

F. GC shall not use Contract Drawings for Shop Drawings and shall provide original Shop Drawings with changes from Contract Drawings clearly indicated.

G. No item shall be marked as "By Others" or similarly.

FIGURE B.2 Submittal procedures specification section—based on MasterSpec template (provided courtesy of Avitru, the developer of AIA MasterSpec).

1.4 PARTIAL SUBMITTALS

A. Design team will not review partial or incomplete Submittals. Complete Submittals for each item are required. Submittals will not be considered official until it is complete in every respect. Delays resulting from partial Submittals are the responsibility of the GC.

1.5 RESUBMITTALS

A. The resubmittal procedure is the same as that specified for initial Submittals. Changes made since the previous Submittal are to be identified.

B. Design team will record time required to review resubmittals after original Submittal and first resubmittal. University retains the right to require the GC to reimburse University for additional fees from the design team for reviewing Submittal more than two (2) times unless the processing results from approved Change Orders causing revisions to previously approved Submittals.

1.6 DISTRIBUTION

A. Duplicate and distribute reproductions of Shop Drawings, product data, samples, and other Submittals which bear the design team's stamp of approval to the Project record documents file, the Project site file, Trade/Subcontractors, Suppliers, other affected Contractors, and other entities requiring the information.

B. Provide each testing and inspection agency one set of approved Submittals for their exclusive use in providing specified quality control testing and inspection services. See §01 04 30.

1.7 SUBMITTAL FORM

FIGURE B.2 (*Continued*)

SUBMITTAL FORM

Submit separate form for each product

Project Name:

Project Code:

TRANSMITTAL A

To: GC ______ Date: ______ Submittal # ______

From: Trade/Subcontractor ______ By: ______ Resubmission ❐

Quantity	Reference #	Title/Description/ Manufacturer	Spec Section Title, Paragraph/Drawing Detail Reference

- ❐ Submitted for Review & Approval
- ❐ Resubmitted for Review & Approval
- ❐ Complies with Contract requirements
- ❐ Will be available to meet construction schedule
- ❐ A/E review time included in construction schedule

Other remarks on above submission:

- ❐ Substitution involved - substitution request attached
- ❐ If substitution involved, submission includes full point by point comparative data or preliminary details
- ❐ Items included in submission will be ordered immediately upon receipt of approval

TRANSMITTAL B

Reviewed, Coordinated, and Approved by GC

To: A/E - Attention ______ Date Received by GC: ______

From: GC ______ Date Transmitted by GC: ______

Remarks on above submission:

One copy retained by sender

TRANSMITTAL C

To: GC ______ Date Received: ______

From: A/E ❐ Other Signed ______ Dated Returned: ______

- ❐ Approved
- ❐ Approved as Noted
- ❐ Disapproved/Resubmit
- ❐ Not Subject to Review
- ❐ Submission Incomplete; Resubmit
- ❐ Received; No Action Required
- ❐ Full point by point comparative data required to complete approval process
- ❐ Provide File Copy with corrections identified
- ❐ Reproducible copies only returned

Remarks on above submission:

One copy retained by sender

TRANSMITTAL

To: Trade/Subcontractor ______ Date Received: ______

From: GC Signed ______ Dated Returned: ______

Copies ❐ University ❐ ❐ ❐ One copy retained by sender

Remarks on above:

END OF SECTION 01 03 30

FIGURE B.2 *(Continued)*

SECTION 31 25 00 - EROSION AND SEDIMENT CONTROL

PART 1 - GENERAL

1.1 RELATED DOCUMENTS

A. Drawings and general provisions of the Contract, including General Conditions, Supplemental General Conditions, and Division 01 Specification Sections, apply to this Section.

1.2 SUMMARY

A. Compliance: This WORK shall be performed in accordance with the erosion and sediment control plan of the construction Drawings and the details provided therein, and as described, detailed and required in the most recent edition of the Virginia Erosion and Sediment Control Handbook.

B. Details: In the event a detailed plan is not shown, CONTRACTOR shall comply with the requirements of the local authority and provide a plan if required by that authority.

1.3 REFERENCES

A. Virginia Erosion and Sediment Control Handbook, latest edition.

1.4 SUBMITTALS

A. Temporary Seed Mixture: Provide written notification as to the temporary seed mixture to be used.

1.5 MAINTENANCE

A. Maintain all erosion and sediment control structures to be utilized during the life of the Project in compliance with the regulations of the Virginia Erosion and Sediment Control Handbook, latest edition, until vegetative cover is acceptable to the inspection personnel and approval acceptance is received.

B. The General Contractor shall maintain all Erosion and Sediment Controls while installing plantings, vegetation, grass, mulch, etc.

PART 2 - PRODUCTS

2.1 STORM DRAIN INLET PROTECTION

A. Sediment deposits shall be removed when deposits reach approximately one-half the design depth of the trap.

PART 3 - EXECUTION

3.1 EROSION AND SEDIMENT CONTROL MEASURES

A. Measures: Silt fence, straw bales, construction entrance, check dams, temporary vegetation, and all other items for erosion and sediment control shall be constructed as directed by the ENGINEER or in the locations shown or designated on the Drawings in accordance with the details provided.

FIGURE B.3 Erosion and sediment control specification section—based on MasterSpec template (provided courtesy of Avitru, the developer of AIA MasterSpec).

B. Schedule: CONTRACTOR shall institute the erosion and sediment control program as a part of clearing and grubbing, and prior to rough grading. The initial program shall include, however, is not limited to, the installation of silt fences, staked straw bales, diversion ditches and/or gravel weirs as shown on erosion and sediment control Drawing at the limits of clearing and grubbing where silt-carrying surface water runoff may be diverted and/or filtered prior to leaving the disturbed area.

C. Temporary Seeding: Establish temporary cover for erosion control by seeding and/or mulching graded areas which may otherwise be exposed for a period greater than 30 days. This should be accomplished as soon as rough grading WORK is done. Begin temporary seeding within 72 hours after earth disturbance.

D. Pipe Outfalls: All pipe outfall areas disturbed by construction shall be protected with non-erodible materials conforming to the Virginia Erosion and Sediment Control Handbook.

E. Maintenance: All siltation and erosion control devices installed during the course of construction shall be maintained in proper working order at all times and shall not be removed until final stabilization of all disturbed areas or at the direction of the ENGINEER.

3.2 CLEANING OF ROADS AND STREETS

A. CONTRACTOR shall maintain a vehicle wash rack or gravel bed at all vehicle egress areas. All vehicles shall be thoroughly cleaned of mud and silt before leaving the construction site to avoid tracking mud and silt onto roads, streets, and highways. In the event that tracking does occur, CONTRACTOR shall immediately clean the street or road of all debris, mud or silt and shall pay all damages resulting therefrom. A daily survey of the condition of the adjacent streets and roads shall be made and recorded in the field log along with daily cleanup of the streets of the tracking from the site onto roads, alleys, parking lots, and highways.

3.3 PROTECTION OF STORMWATER SYSTEMS

A. Stormwater structures which will receive runoff from the construction shall be protected from the buildup of mud or silt as outlined by the Virginia Erosion and Sediment Control Handbook or as directed by ENGINEER.

END OF SECTION 31 25 00

FIGURE B.3 *(Continued)*

Appendix C

Cost Estimate Form

Project Name ______________________________

Prepared By ______________________ Date ______________

Checked By ______________________ Date ______________

	Item	Quantity	Unit	Unit Price	Cost
A.	**Clearing and Grading**				
	1. Clearing and Grubbing	______	Acre	______	______
	2. Disposal	______	L.S.	______	______
	3. Excavation (cut)	______	C.Y.	______	______
	4. Embankment (fill)	______	C.Y.	______	______
	5. Borrow	______	C.Y.	______	______
	6. Spoil/Disposal	______	C.Y.	______	______
	7. Topsoil Removal/Storage	______	C.Y.	______	______
	8. ______________	______	____	______	______
		Sub-Total for Clearing and Grading			______
B.	**Erosion/Sediment Control**				
	1. Earth Berm	______	L.F.	______	______
	2. Perimeter Dike	______	L.F.	______	______
	3. Interceptor Dike	______	L.F.	______	______
	4. Filter Fabric Fence	______	L.F.	______	______
	5. Gravel Filters: (Excl. filters with basins)	______	C.Y.	______	______
	6. Inlet Sediment Traps	______	Each	______	______
	7. Sod	______	S.Y.	______	______
	8. Seeding (disturbed area minus pavements and buildings)	______	S.Y.	______	______
	9. Sediment Basins (including filters)	______	Each	______	______
	10. Wash Rack (incl. water service)	______	Each	______	______
	11. ______________	______	____	______	______
		Sub-Total For Erosion/Sedimtent Control			______
C.	**Streets**				
	1. Bituminous Concrete Surface:				
	1"	______	S.Y.	______	______
	2"	______	S.Y.	______	______
	2. Base Course: 6" Aggregate	______	S.Y	______	______
	3" Bit. Conc.	______	S.Y.	______	______
	3. Aggregate Sub-base	______	C.Y.	______	______
	4. Cement Treated Aggregate	______	C.Y.	______	______
	5. Cement Treated Sub-grade	______	C.Y.	______	______
	6. Surface Treatment	______	S.Y.	______	______
	7. Gravel Shoulder	______	S.Y.	______	______
	8. Curb & Gutter	______	L.F.	______	______
	9. Header Curb	______	L.F.	______	______
	10. Median Curb	______	L.F.	______	______
	11. Curb Cut Ramps	______	Each	______	______
	12. Commercial Entrance	______	L.F.	______	______
	13. Driveway Entrance	______	Each	______	______
	14. Guard Rail	______	L.F.	______	______
	15. Traffic Barricade	______	Each		

FIGURE C.1 Cost estimate form.

	16. Edge Delineators			Each		
	17. Sidewalk (4' wide, concrete)			L.F.		
	18. Rock Excavation			C.Y.		
	19. Soil Testing (Road)			L.F.		
	20. ________					
				Sub-Total for Streets		
D.	**Storm Sewer**					
	1. Easement Clearing & Restoration			L.F.		
	2. Pipe RCCP, Classes II-IV:					
		15"		L.F.		
		18"		L.F.		
		21"		L.F.		
		24"		L.F.		
		27"		L.F.		
		30"		L.F.		
		33"		L.F.		
		36"		L.F.		
		42"		L.F.		
		48"		L.F.		
				L.F.		
	3. Pipe, CMP:	15"		L.F.		
		18"		L.F.		
		21"		L.F.		
		24"		L.F.		
		27"		L.F.		
		30"		L.F.		
		33"		L.F.		
		36"		L.F.		
				L.F.		
	4. Endwall	15"		Each		
		18"		Each		
		21"		Each		
		24"		Each		
		27"		Each		
		30"		Each		
		33"		Each		
		36"		Each		
		____		Each.		
	5. End Section	15"		Each		
		18"		Each		
		21"		Each		
		24"		Each		
		27"		Each		
		30"		Each		
		33"		Each		
		36"		Each		
		42"		Each		
		48"		Each		
		____		Each		
	6. Endwall			Each		
				Each		

Figure C.1 *(Continued)*

7. Curb Inlet	______	Each	______	______
8. Other Inlets	______	Each	______	______
	______	Each	______	______
9. Junction Box JB-1 Pipe size: ____	______	Each	______	______
____	______	Each	______	______
10. Manhole	______	Each	______	______
11. Modified Structure	______	Each	______	______
12. Connect to Existing Structure	______	Each	______	______
13. Remove Existing Structure	______	Each	______	______
14. Pavement Restoration	______	S.Y.	______	______
15. Curtain Wall	______	C.Y.	______	______
16. Concrete Cradle/Encasement	______	L.F.	______	______
17. Concrete Anchors	______	C.Y.	______	______
18. Paved Ditch	______	C.Y.	______	______
19. Sod Ditch	______	S.Y.	______	______
20. Rip Rap: Class I Ungrouted	______	C.Y.	______	______
Class I Grouted	______	C.Y.	______	______
Class II	______	C.Y.	______	______
21. Piling	______	L.S.	______	______
22. Box Culvert (compute Cost on Box Culvert Computation Form)	______	L.S.	______	______
23. Underdrains	______	L.F.	______	______
24. ____________	______	____	______	______
		Sub-Total for Storm Sewer		______

E. Sanitary Sewer

1. Easement Clearing & Restoration	______	L.F.	______	______
2. Pipe,(Including 0-8' Trenching and Select Fill): 8"	______	L.F.	______	______
10"	______	L.F.	______	______
3. Pipe, RCCP (Including 0-8' Trenching and Select Fill): 8"	______	L.F.	______	______
10"	______	L.F.	______	______
4. Pipe, Ductile Iron (Including 0-8' Trenching and Select Fill): 8"	______	L.F.	______	______
10"	______	L.F.	______	______
5. Extra Trenching (Not includng Pipe): 8'-12'	______	L.F.	______	______
12-16'	______	L.F.	______	______
16'+	______	L.F.	______	______
6. Manhole (All Manholes) 0'-8'	______	Each	______	______
(Depth over 8') 8'+	______	V.F.	______	______
7. Drop Connection	______	V.F.	______	______
8. Concrete Encasement	______	L.F.	______	______
9. Concrete Cradle	______	L.F.	______	______
10. Concrete Anchors	______	C.Y.	______	______
11. Rock Excavation	______	C.Y.	______	______
12. Sheeting and Shoring	______	L.S.	______	______
13. Pavement Replacement	______	L.F.	______	______
14. Laterals: 4"	______	L.F.	______	______
5"	______	L.F.	______	______
15. Risers	______	V.F.	______	______

FIGURE C.1 *(Continued)*

	16. "T" or "Y" Connection:				
	8" x 4"	________	Each	________	________
	8" x 5"	________	Each	________	________
	10" x 4"	________	Each	________	________
	17. Lateral Connection to Manhole	________	Each	________	________
	18. Connect to Existing Structure	________	Each	________	________
	19. Aerial Pipe Supports	________	L.S.	________	________
	20. Boring or Jacking	________	L.F.	________	________
	21. Select Fill Material	________	C.Y.	________	________
	22. ____________________	________	____	________	________
			Sub-Total for Sanitary Sewer		________
F.	**Water Line**				
	1. Easement Clearing & Restoration	________	L.F.	________	________
	2. Pipe, DIP: 3"	________	L.F.	________	________
	4"	________	L.F.	________	________
	6"	________	L.F.	________	________
	8"	________	L.F.	________	________
	10"	________	L.F.	________	________
		________	L.F.	________	________
	3. Valve 3"	________	Each	________	________
	4"	________	Each	________	________
	6"	________	Each	________	________
	8"	________	Each	________	________
	10"	________	Each	________	________
		________	Each	________	________
	4. Fittings (Tees, Bends Crosses, Reducers, and Offsets)	________	LB.	________	________
	5. Copper Water Service (Includes Curb Stop) Note: Tap fee may cover this item	________	L.F.	________	________
	6. Wet Tap	________	Each	________	________
	7. Fire Hydrant	________	Each	________	________
	8. Blow-off	________	Each	________	________
	9. Air Release	________	Each	________	________
	10. Rock Excavation	________	C.Y.	________	________
	11. Boring and/or Jacking	________	L.F.	________	________
	12. Pavement Restoration	________	L.F.	________	________
	13. Shoulder Restoration	________	L.F.	________	________
	14. Concrete Anchor/Encasement	________	C.Y.	________	________
	15. Concrete Cradle	________	L.F.	________	________
	16. Relocate Existing Hydrant	________	Each	________	________
	17. Sheeting and Shoring	________	L.F.	________	________
	18. Select Fill Material	________	C.Y.	________	________
	19. ____________________	________	____	________	________
			Sub-Total for Water		________
G.	**Miscellany**				
	1. Septic Field	________	L.S.	________	________
	2. Pump	________	Each	________	________
	3. Monuments and Pipes	________	L.S.	________	________
	4. Monuments and Pipes	________	Lot	________	________
	5. Walkways and Trails: 4' Asphalt	________	L.F.	________	________
	8' Asphalt	________	L.F.	________	________
	4' Gravel	________	L.F.	________	________
	4' Woodchip	________	L.F.	________	________

Figure C.1 *(Continued)*

6. Landscaping	________	L.S.	________	________
7. Recreational Amenities	________	L.S.	________	________
8. Screening	________	L.S.	________	________
9. Fencing	________	L.S.	________	________
10. Retaining Wall	________	L.S.	________	________
11. Well	________	Each	________	________
12. Street Sign	________	Each	________	________
13. Wheel Stops	________	Each	________	________
14. ________________	________	_____	________	________
		Sub-Total for Miscellaney		________

H. <u>Total Cost</u>

		Cost
1. Sub-Totals:	Clearing and Grading	____________
	Siltation/Erosion Control	____________
	Streets	____________
	Storm Sewer	____________
	Sanitary Sewer	____________
	Water	____________
	Miscellany	____________
	GRAND SUB-TOTAL	____________
2. Engineering & Contingencies (%)		____________
	TOTAL COST	____________

Note: Cost Data Taken from available records.

FIGURE C.1 *(Continued)*

Bibliography

American Congress on Surveying and Mapping, and American Society of Civil Engineers. Definitions of Surveying and Associated Terms. Revised 1978. Reprint. Bethesda, Maryland: American Congress on Surveying and Mapping, 1989. Revised 2005.

American Institute of Constructors Construction Certification Commission (AICCCC). *Associate Constructor Study Guide: Level 1—Construction Fundamentals*, revised and printed May 11, 2005. Alexandria, VA: American Institute of Constructors (AIC).

Anderson, James M., and Edward M. Mikhail. 1998. *Surveying: Theory and Practice*, 7th ed. New York: WCB/McGraw-Hill.

Balmori, Diana, and Gaboury Benoit. 2007. *Land and Natural Development (LAND) Code: Guidelines for Sustainable Land Development*. New York: John Wiley & Sons.

Barnes, Martin. 1988. "Construction Project Management." *International Journal of Project Management* 6(2):69–79.

Buckner, R. Ben. 2000. *Land Survey Review Manual*, 3rd ed. Michigan: Ann Arbor Press.

Building Commissioning Association. 2011. *New Construction Building Commissioning Best Practice*, December 8, pp. 14–21, https://www.bcxa.org/wp-content/pdf/BCA-Best-Practices-Commissioning-New-Construction.pdf.

Cheves, Marc. 1990. "Making Data Collection Work." *P.O.B.—Point of Beginning* 15(6):54, 58, 60.

Colley, Barbara C. 1998. *Practical Manual of Land Development*, 3rd ed. New York: McGraw-Hill Professional.

Connecticut Department of Transportation (CTDOT). 2017. "Project Documentation." In *Contract Administration*. Vol. 1 of *Construction Manual*, ver. 3.0. CTDOT (January): 1-3.1–27. https://www.ct.gov/dot/lib/dot/documents/dconstruction/construction_manual/CM_ver_3.0.pdf.

Construction Industry Institute (CII). "IR102-2—Partnering Toolkit." Accessed April 30, 2018. https://www.construction-institute.org/resources/knowledgebase/best-practices/partnering/topics/rt-018/pubs/sp17-1.

Construction Management Association of America (CMAA). *Construction Management Standards of Practice*, 2015 ed. McLean, VA: CMAA, 2015.

Cox, Tom, Michael Kenig, Markku Allison, Suzanna Wight Kelley, American Institute of Architects, and Michael Stark. *Primer on Project Delivery*, 2nd ed. Washington, DC: The American Institute of Architects; Arlington, VA: The Associated General Contractors of America, 2011. https://www.agc.org/sites/default/files/Files/Programs%20%26%20Industry%20Relations/AIA-AGC_Primer_on_Project_Delivery_2nd_Edition-FINAL.pdf

Crown Copyright. 2008. "Communities and Local Government Publications." *Performance Testing of Buildings*, pp. 9–42, https://www.bipsolutions.com/docstore/pdf/20548.pdf.

David Jensen Associates, Inc. 1984. *How to Win at the Zoning Table*. Washington, DC: National Association of Home Builders.

Davis, Raymond, Francis Foote, and Joe Kelly. 1968. *Surveying: Theory and Practice*, 5th ed. New York: McGraw-Hill.

De Marco, Alberto. 2011. "Planning and Scheduling." In *Project Management for Facility Constructions*. Berlin: Springer, pp. 89–116.

Denny, Milton. Noteworthy Field Notes. 1999. *P.O.B.—Point of Beginning* 25(2):27–28.

Dewberry & Davis. 1981. *Procedure & Checklist Manual*. In-house guidelines. Fairfax, VA: Dewberry & Davis.

Dewberry & Davis Institute. 1984. *Party Chief Training Manual*. In-house training text. Fairfax, VA: Dewberry & Davis.

Engineers and Surveyors Institute. 1988. *Improving Preparation, Review & Approval of Subdivision & Site Plans in Fairfax County*. Fairfax, VA: Engineers and Surveyors Institute.

Engineers Joint Contracts Document Committee (EJCDC), 2018. "Court Decisions." Accessed April 30, 2018. https://www.ejcdc.org/court-decisions/.

Engineers Joint Contracts Document Committee. 2013. *Bid Form for Construction Contracts* (EJCDC C-410). Alexandria, VA: National Society of Professional Engineers; Washington, DC: American Council of Engineering Companies; Reston, VA: American Society of Civil Engineers.

Engineers Joint Contracts Document Committee. 2013. *Guide to the Preparation of Supplementary Conditions* (EJCDC C-800). Alexandria, VA: National Society of Professional Engineers; Washington, DC: American Council of Engineering Companies; Reston, VA: American Society of Civil Engineers.

Engineers Joint Contracts Document Committee. 2014. *Standard Form Agreement between Owner and Engineer for Professional* Services (EJCDC E-500). Alexandria, VA: National Society of Professional Engineers; Washington, DC: American Council of Engineering Companies; Reston, VA: American Society of Civil Engineers.

Engineers Joint Contracts Document Committee. 2013. *Standard General Conditions of the Construction Contract* (EJCDC C-700). Alexandria, VA: National Society of Professional Engineers; Washington, DC: American Council of Engineering Companies; Reston, VA: American Society of Civil Engineers.

Engineers Joint Contracts Document Committee. 2013. *Suggested Instructions to Bidders for Construction Contracts* (EJCDC C-200). Alexandria, VA: National Society of Professional Engineers; Washington, DC: American Council of Engineering Companies; Reston, VA: American Society of Civil Engineers.

Federal Highway Administration (FHWA). 2004. "Construction Program Management and Inspection Guide." *US Department of Transportation*. https://www.fhwa.dot.gov/construction/cpmi04.pdf.

Grant, Eugene L., W. Grant Ireson, and Richard S. Leavenworth. 1990. *Principles of Engineering Economy*, 8th ed. New York: John Wiley & Sons.

Halpin, Daniel W., Bolivar A. Senior, and Gunnar Lucko. 2017. *Construction Management*, 5th ed. Hoboken, NJ: John Wiley & Sons.

Hancher, Donn. 2003. "Construction Planning and Scheduling." In *Civil Engineering Handbook*, 2nd ed., edited by W.F. Chen and J.Y. Richard Liew. Boca Raton, FL: CRC Press, LLC, pp. 2-1–25.

Herubin, Charles. 1982. *Principles of Surveying*, 3rd ed. Reston, VA: Reston Publishing Company.

Institute for Participatory Management & Planning. 1986. *Citizen Participation Handbook*. Laramie, Wyoming: Institute for Participatory Management & Planning.

International City Management Association. 1979. *The Practice of Local Government Planning*. Washington, DC: International City Management Association.

Jha, K.N., and K.C. Iyer. 2006. "Critical Factors Affecting Quality Performance in Construction Projects." *Total Quality Management & Business Excellence* 17(9): 1155–1170. doi:10.1080/14783360600750444.

Kelly, John, and Steven Male. 2005. *Value Management in Design and Construction: The Economic Management of Projects*. London: Routledge, pp. 1–4. Taylor and Francis e-Library.

Klein, Gabe, Ronaldo Nicholson, Greer Gillis, Said Cherifi, and Mesfin Lakew. 2010. "Construction Management Manual." *District of Columbia Department of Transportation*, pp. 1–41. https://ddot.dc.gov/sites/default/files/dc/sites/ddot/publication/attachments/ddot_construction_management_manual_rev07-01-2010.pdf.

Knutson, Kraig, Clifford Schexnayder, Christine Fiori, and Richard Mayo. 2009. *Construction Management Fundamentals*, 2nd ed. New York: McGraw-Hill.

Lakshmi, R. 2015. "Quality Control and Quality Assurance in Building Construction." In *National Conference of Research Advances in Communication, Computation, Electrical Science and Structures*, pp. 29–37.

Los Alamos National Laboratory (LANL). 2002. "Commissioning the Building." In *LANL Sustainable Design Guide*. LANL, pp. 177–185. https://www.energy.gov/sites/prod/files/2013/12/f5/sustainable_guide_ch9.pdf.

Moffitt, Frances, and Harry Bouchard. 1982. *Surveying*, 7th ed. New York: Intext Educational Publishers.

Munjy, Riadh, et al. 1989. "Total Station Survey System (TSSS) Software. Surveying and Mapping." *Journal of American Congress on Surveying and Mapping* 49(4):31, 34-35.

National Association of Home Builders. 1978. *Subdivision Regulation Handbook*. Washington, DC: National Association of Home Builders.

Pafford, F. William. 1962. *Handbook of Survey Note-keeping*. New York: John Wiley & Sons.

Park, William R., and Dale E. Jackson. 1984. *Cost Engineering Analysis: A Guide to Economic Evaluation of Engineering Projects*. New York: John Wiley & Sons.

Peña-Mora, Feniosky, Carlos E. Sosa, and Sean D. McCone. 2002. *Introduction to Construction Dispute Resolution*. Upper Saddle River, NJ: Prentice-Hall.

Peurifoy, Robert, Clifford Schexnayder, Aviad Shapira, and Robert Schmitt. 2011. *Construction Planning, Equipment, and Methods*, 8th ed. New York: McGraw-Hill.

Pulaski, Michael H., and Michael J. Horman. 2005. "Continuous Value Enhancement Process." *Journal of Construction Engineering and Management* 131(12): 1274–1282. https://ascelibrary.org/doi/10.1061/%28ASCE%290733-9364%282005%29131%3A12%281274%29.

Root, Ray B., LLS. Personal Communication. May 1, 2001.

Roy, Rajat, Margaret Low, and John Waller. 2005. "Documentation, Standardization and Improvement of the Construction Process in House Building." *Construction Management and Economics* 23(1):57–67. https://doi.org/10.1080/0144619042000287787.

Shrestha, Ramesh. 1990. "Formats and Specifications for an Electronic Field Book. Surveying and Land Information Systems." *Journal of American Congress on Surveying and Mapping* 50(3):215–224.

Spear, D.F. 1969. "S.P.C.L.—A New Approach to Value Engineering." *Value Engineering* 1(5): 273–276.

Transportation Research Board (TRB). 2013. "Glossary of Transportation Construction Quality Assurance Terms," 6th ed. *Transportation Research Circular*, Number E-C173. Washington, DC: TRB, 10.

Urban Land Institute. 1985. *Working with the Community: A Developer's Guide*. Washington, DC: The Urban Land Institute.

US General Services Administration (GSA). 2005. *The Building Commissioning Guide*, pp. 49-53, https://www.gsa.gov/cdnstatic/BCG_3_30_Final_R2-x221_0Z5RDZ-i34K-pR.pdf.

Virginia Association of Surveyors, Inc. 1993. *Standards of Practice for Professional Land Surveyors in the Commonwealth of Virginia*, 3rd ed. Richmond, VA: VAS, Inc.

Washington State Department of Transportation (WSDOT). 2018. "Documentation." In *Construction Manual*. M41-01.32. WSDOT, 2–30. https://www.wsdot.wa.gov/publications/manuals/fulltext/M41-01/Construction.pdf.

Whitney, Donna, President, Geomatic Science, Inc. Interview by Lisa Rauenzahn. August 2007.

Whitney, Ted, Professional Land Surveyor, Geomatic Science, Inc. Interview by Lisa Rauenzahn. August 2007.

Wigfield, George, Professional Land Surveyor, Dewberry & Davis LLC. Interview by Sam Dougherty. Lanham, MD. August 2007.

Wolf, Paul R., and Charles D. Ghilani. 2006. *Elementary Surveying: An Introduction to Geomatics*, 11th ed. Upper Saddle River, NJ: Pearson Prentice Hall.

Zhang, Xueqing, Xiaoming Mao, and Simaan M. AbouRizk. 2009. "Developing a Knowledge Management System Practices in the Construction Industry." *Automation in Construction* 18:777–789. doi:10.1016/j.autcon.2009.03.004.

INDEX

www.ingramcontent.com/pod-product-compliance
Lightning Source LLC
LaVergne TN
LVHW082227200926
848484LV00008B/2312